Meta-Analysis, Decision Analysis, and Cost-Effectiveness Analysis

Monographs in Epidemiology and Biostatistics

Edited by Jennifer L. Kelsey, Michael G. Marmot,
Paul D. Stolley, Martin P. Vessey

MONOGRAPHS IN EPIDEMIOLOGY AND BIOSTATISTICS · VOLUME 24

Meta-Analysis, Decision Analysis, and Cost-Effectiveness Analysis

Methods for Quantitative Synthesis in Medicine

Diana B. Petitti

New York   Oxford
OXFORD UNIVERSITY PRESS
1994

Oxford University Press

Oxford   New York   Toronto
Delhi   Bombay   Calcutta   Madras   Karachi
Kuala Lumpur   Singapore   Hong Kong   Tokyo
Nairobi   Dar es Salaam   Cape Town
Melbourne   Auckland   Madrid

and associated companies in
Berlin   Ibadan

Published by Oxford University Press, Inc.,
200 Madison Avenue, New York, New York 10016

Oxford is a registered trademark of Oxford University Press

Library of Congress Cataloging-in-Publication Data
Petitti, Diana B.
Meta-analysis, decision analysis, and cost-effectiveness analysis :
methods for quantitative synthesis in medicine / Diana B. Petitti.
p. cm. — (Monographs in epidemiology and biostatistics ; 23)
Includes bibliographical references and index.
ISBN 0-19-507334-7
1. Medicine—Decision making.   2. Statistical decision.   3. Meta-
analysis.   4. Cost effectiveness.   I. Title.   II. Series:
Monographs in epidemiology and biostatistics ; v. 23.
[DNLM: 1. Meta-Analysis.   2. Decision Support Techniques.   3. Cost-
Benefit Analysis.   W1 MO567LT v. 23 1994 / WA 950 P491m 1994]
R723.5.P48   1994
614.4′072—dc20
DNLM/DLC
for Library of Congress   93-30258

9 8 7 6 5 4 3 2

Printed in the United States of America
on acid-free paper

Preface

Meta-analysis, decision analysis, and cost-effectiveness analysis are conceptually related quantitative methods for combining information to arrive at a summary conclusion. Development of each of the three methods grew out of the need to resolve uncertainty: for meta-analysis, uncertainty about the medical literature; for decision analysis, uncertainty about management of clinical problems; and for cost-effectiveness analysis, uncertainty about how best to allocate resources. All play a large and increasingly important role in formulation of policy recommendations in medicine. Their use in summarizing information from the burgeoning, often contradictory literature of medicine has become widespread.

I decided to write a book describing how to do these three kinds of research in the winter of 1990. At that time, I was working with a number of students and fellows who wanted experience doing research but who had no time to carry out meaningful studies that involved collecting data from human subjects. I had previously identified several important questions that I thought might be addressed with meta-analysis, decision analysis, or cost-effectiveness analysis, and I recruited students to work on these projects with me over the spring and summer.

I soon discovered that the seductive simplicity of the results of published studies that used the three methods belied enormous complexity in their proper conduct. It became apparent in working with students that successful use of any of the methods required a great deal of knowledge about research study design, practical skills in data collection, a firm understanding of statistics, and an ability to critically evaluate and recognize the limitations inherent in each method. Much to my dismay, no generalized discussion of any of the three methods written at a level appropriate for students or fellows was available. What had been written tended to focus on the quantitative aspects of each method to the exclusion of discussion of the practical aspects of information retrieval and interpretation.

More important, there had never been an attempt to unify the three topics, although their conceptual similarity as methods that "synthesize" information quantitatively seems obvious. The three methods are considered together in this book in the belief that a scientist interested in one is also likely to be eventually interested in all three. In addition, meta-analysis can often enhance estimation of probabilities in a decision analysis, and constructing a decision tree is the first step in a cost-effectiveness analysis.

This book describes how to design, conduct, analyze, and interpret these three types of synthetic studies. The book describes in detail how to collect the data necessary to do the studies. Basic statistical methods for meta-analysis and quantitative methods necessary to do decision analysis are presented. Information on how to measure cost is provided. The book includes many published examples of analyses using the three techniques as well as some unpublished data. Examples of analyses that yield incorrect or problematic conclusions are also given, as there is often more to be learned from failure than from success. The limitations of the methods are described fully, and, I hope, honestly.

The book tries to present sufficient detail on the practical aspects of the application of the methods so that a novice could do a simple analysis knowing nothing more than what is in the book. The book attempts not to repeat advanced material on decision analysis that is covered well in textbooks by Weinstein and Fineberg *(Clinical Decision Analysis)* and Sox et al. *(Medical Decision Making)*. Cost-effectiveness analysis is covered at a level suitable for many physicians, but it will not be useful to economists or experts in cost-effectiveness analysis.

The reader who has taken an introductory course in both epidemiology and biostatistics should be prepared to read the basic text. Some of the material on statistical analysis of meta-analytic data may require more advanced preparation. Even the reader with little direct experience in carrying out studies using these methods will profit from reading the discussion of the limitations of the methods. In addition, policymakers without direct research experience will learn how better to utilize data from the three types of studies.

In a perfect world, every important topic in medicine would be studied using a randomized clinical trial; a single definitive study would always be available; when interventions were known to work, they would be available to everyone. The reality of the scientific and fiscal world in which we live is far from this ideal. It is not possible to study all medical treatments in experiments. Single studies are almost never definitive. We cannot afford to pay for everything that might be done for patients. Given these realities, the need for synthesis of information to evaluate effectiveness and to evaluate costs is irrefutable. My hope is that this book will help foster rational decision making in medicine based on systematic use of the information we already have.

San Francisco   D.B.P.
May 1993

Contents

Meta-Analysis, Decision Analysis, and Cost-Effectiveness Analysis

1

Introduction

The rapid accumulation of medical information makes decisions about patient management more and more difficult for the practicing physician. Increasing numbers of medical technologies and the availability of several different technologies for the management of the same condition greatly complicate decision making at both the individual and the policy level. The need to develop guidelines for clinical practice and to set funding priorities for medical interventions is pressing, but the volume and the complexity of medical information thwarts attempts to make firm recommendations about what should and should not be done and what should and should not be paid for. It is in this context that the three related research methods—meta-analysis, decision analysis, and cost-effectiveness analysis—have been developed and have gained popularity. The overall objective of this book is to show how these research methods can be used to meet the growing challenge of making sense of what is known in order to maximize the utility of medical knowledge.

Section 1.1 presents three problems that illustrate the need for these research methods; it describes the reasons for considering the three methods in a single text. Section 1.2 defines the three methods and shows their application to the three illustrative problems. Section 1.3 describes the history of the application of each method in medicine. Section 1.4 gives an example of how the methods can be used together to comprehensively address a single medical problem, foreshadowing the potential use of the methods. Section 1.5 briefly describes the organization of the book.

1.1 THREE ILLUSTRATIVE PROBLEMS

A physician must decide whether to recommend that patients with transient ischemic attacks use an antiplatelet drug to prevent stroke.

A committee charged with developing guidelines on the management of patients with HIV disease must decide whether to recommend isoniazid as preventive therapy in HIV-infected intravenous drug users.

The Health Care Financing Administration must decide whether Medicare will pay for beneficiaries to receive pneumococcal vaccine.

Each of these problems is a real problem, either past or present. Each problem requires a policy solution—a solution for the patients of one physician, for a group of patients, or for an organization. Like almost all other contemporary medical problems, these three problems cannot be addressed by looking at only one piece of information. There are many different studies of antiplatelet drugs in patients with transient ischemic attacks, and no single study is definitive. A decision to recommend isoniazid must consider the toxicity of isoniazid, the likelihood of tuberculosis in an HIV-infected intravenous drug user, and the consequences of tuberculosis. Deciding to pay for pneumococcal vaccine must take into account the effectiveness of the vaccine and its cost in relation to competing demands for Medicare resources. Addressing each problem requires the *synthesis* of information.

Synthesis is the bringing together of parts or elements to form a whole (*Webster's New Collegiate Dictionary* 1982). Meta-analysis, decision analysis, and cost-effectiveness analysis have in common that they synthesize knowledge. Each method takes parts of the medical literature or of clinical experience and creates from this information a whole answer to a defined problem.

Each of the three methods is quantitative, using statistical and numerical analysis to create the whole. Each of the methods aims to resolve uncertainty. For meta-analysis, the uncertainty is in what conclusions to draw from a body of research studies on the same topic. For decision analysis, the uncertainty is in what to recommend for a single patient or for a group of similar patients. For cost-effectiveness analysis, the uncertainty is in whether to pay, or how much to pay, for a medical procedure, treatment, or service. Each method aims to facilitate decision making, particularly decision making at the policy level, and each has come to play a prominent role in the policy arena.

1.2 DEFINITIONS

1.2.1 Meta-Analysis

Meta-analysis is a quantitative approach for systematically combining the results of previous research in order to arrive at conclusions about the body of research. Studies of a topic are first systematically identified. Criteria for including and excluding studies are defined, and data from the eligible studies are abstracted. Last, the data are combined statistically, yielding a quantitative estimate of the size of the effect of treatment and a test of homogeneity in the estimate of effect size.

EXAMPLE: The question of whether to prescribe an antiplatelet drug for patients with transient ischemic attacks to prevent stroke was controversial in 1988. At that time, many randomized trials of antiplatelet drugs to treat patients with cerebrovascular disease had been completed, but the studies were variable in quality and their results were contradictory. A meta-analysis of these studies by the Antiplatelet Trialists' Collaboration (1988) found a highly significant 22% reduction in the estimated relative risk of stroke, myocardial infarction, and vascular death in patients with cerebrovascular disease who were treated with an antiplatelet drug. This meta-analysis strongly supports a decision to prescribe an antiplatelet drug for patients with transient ischemic attacks.

1.2.2 Decision Analysis

Decision analysis is a quantitative approach that assesses the relative value of different decision options (Weinstein and Fineberg 1980; Pauker and Kassirer 1987). The information from decision analysis is used to decide how to manage an individual patient[1] or to formulate policy recommendations about a group of similar patients. Decision analysis begins by systematically breaking a problem down into its components and creating a decision tree to represent the components and the decision options. Outcomes of the decision options are defined. Uncertainties in the components are identified, and review of the medical literature and expert opinion are used to estimate probabilities for the uncertainties. Values of the outcomes are measured or inferred.[2] Last, the decision tree is analyzed using statistically based methods. Decision analysis yields an estimate of the net value of the different decision options in relation to each other.

EXAMPLE: In 1991, the question of whether to treat HIV-infected intravenous drug users with isoniazid to prevent active tuberculosis was controversial. These patients are more likely than uninfected persons to be anergic, and the tuberculin skin test is not accurate for this reason. They are also at very high risk of developing tuberculosis both because they use intravenous drugs and because they are infected with HIV. A decision analysis by Jordan et al. (1991) showed that life expectancy would be increased by isoniazid prophylaxis in black and white men and in white women regardless of the results of tuberculin skin tests; life expectancy in black women with negative tuberculin tests would be decreased because of the greater risk of toxicity from isoniazid in black women and because of the lower likelihood of tuberculosis. A committee charged with the task of making recommendations about isoniazid prophylaxis in HIV-infected intravenous drug users could use the results of the decision analysis to recommend prophylaxis in all white and black men who are HIV-infected and intravenous drug users without doing tuberculin skin testing. The recommendation for black women should be to have a tuberculin skin test and to base the decision to treat prophylactically on the results of the test.

1.2.3   Cost-Effectiveness Analysis

Cost-effectiveness analysis compares decision options in terms of their monetary cost. In medical applications, cost-effectiveness analysis usually begins by carrying out a decision analysis. Data on the costs of the decision options are collected, and the costs of the decision options are compared.

The terms "cost effective" and "cost-effectiveness analysis" are frequently misused in medicine (Doubilet, Weinstein, McNeil 1986). They are properly used only when formal analysis comparing the cost of alternative strategies has been done. Cost-effectiveness analysis is different from cost-benefit analysis, but analyses of one type are frequently mislabeled as being of the other type and the words are sometimes used interchangeably (Warner and Luce 1982). In cost-benefit analysis, all of the consequences of the decision options are valued in monetary terms. In cost-effectiveness analysis, at least some of the consequences of the decision options are valued in nonmonetary terms, such as lives saved, years of life saved, or disability avoided.

EXAMPLE: Sisk and Riegelman (1986) did a cost-effectiveness analysis of vaccination against pneumococcal pneumonia in persons over 65 from the point of view of Medicare.[3] They estimated that the net Medicare expenditures for pneumococcal vaccination of the elderly would be $4400 to $8300 (1983 dollars) per year of healthy life gained. They also estimated that vaccination would save money for the Medicare program if it were administered in a public program, where the cost per vaccination would be low. The analysis supports Medicare coverage of pneumococcal vaccine. It suggests that a public program to administer the vaccine would be the most efficient and financially desirable method for vaccinating the elderly.

1.3   HISTORICAL PERSPECTIVE

Attempts to synthesize knowledge are very old. Narrative reviews of the scientific literature have been around for as long as there has been a scientific literature. Quantitative approaches to the synthesis of information appear to be mostly twentieth-century inventions. Over the last two decades, quantitative synthesis has become more popular and more influential. To illustrate, Table 1-1 shows the number of articles using these methods published in four influential general medical journals in successive five-year intervals from 1971 through 1990. Taken together, the growth in publications of articles using one of these three methods has been exponential since 1971.

1.3.1   Meta-Analysis

The term "meta-analysis" was coined by Glass in 1976 from the Greek prefix "meta," which means "transcending," and the root, analysis. Glass sought to dis-

Table 1-1 Number of articles using meta-analysis, decision analysis, or cost-effectiveness analysis published in four influential English-language journals[a]

Year	Meta-Analysis	Decision Analysis	Cost-Effectiveness Analysis	Any
1971–1975	0	4	4	8
1976–1980	0	10	10	20
1981–1985	1	19	20	40
1986–1990	32	20	26	78

[a] *Annals of Internal Medicine, Journal of the American Medical Association, Lancet, New England Journal of Medicine.*

tinguish meta-analysis from primary analysis—the original analysis of data from a research study—and secondary analysis—the reanalysis of data to answer new research questions (Glass 1976; Glass, McGaw, Smith 1981). However, the statistical combination of data from multiple studies of the same topic began long before a word for it was coined. In the 1930s, Tippett (1931), Fisher (1932), Cochran (1937) and Pearson (1938), each described statistical techniques for combining data from different studies, and examples of use of these techniques in the agricultural literature of the 1930s are numerous.

The need for techniques to combine data from many studies of the same topic became especially acute in the social sciences in the mid-1970s, when literally hundreds of studies existed for some topics. Development of meta-analysis was driven by the perception that narrative literature reviews were selective in their inclusion of studies and subjective in their weighting of studies.

During the late 1970s and early 1980s, social scientists, including Rosenthal (1978), Glass, McGaw, and Smith (1981), Hedges (1982, 1983), Hunter, Schmidt and Jackson (1982), Light (1983), and Light and Pillemar (1984), popularized meta-analysis and further developed the statistical methods for its application. Equally important, they expanded the goal of studies that combine data to include the attempt to systematically identify the studies to be combined, and they made estimation of effect size, not just statistical significance, a primary aim of meta-analysis.

The first use of meta-analysis in medicine is difficult to date precisely, but its use in medicine quickly followed its popularization in the social sciences. In the late 1980s, descriptions of the method of meta-analysis appeared almost simultaneously in three influential general medical journals, the *New England Journal of Medicine, Lancet,* and the *Annals of Internal Medicine* (L'Abbe, Detsky, O'Rourke 1987; Sacks et al. 1987; Bulpitt 1988). More widespread use of meta-analysis in medicine has coincided with the increasing focus of medical research on the randomized clinical trial, and it has undoubtedly benefited from the rising level of concern about the interpretation of small and individually inconclusive clinical trials. Use of meta-analysis is not confined to synthesis of information from experimental studies, however. A large number of studies that involve the

Table 1-2   Topics and journals of publications for some meta-analyses of nonexperimental data that were published in 1990–1992

Topic	Journal of Publication
Physical activity and coronary heart disease	*Journal of the American Board of Family Practice*
Chlorination of water and cancer	*American Journal of Public Health*
Accuracy of computed tomography, magnetic resonance imaging, and myelography for spinal stenosis	*American Journal of Roentgenology*
Estrogen excretion in women with breast cancer	*Cancer*
Atrial natriuretic factor in normal and hypertensive persons	*American Journal of Hypertension*
Health risks of work in the nuclear industry	*Occupational Medicine*
Predictors of recurrent febrile seizures	*Journal of Pediatrics*
Outcome in low birth weight infants	*Journal of Pediatrics*
Secular trend in the sensitivity of the chest x-ray for hypersensitivity pneumonitis	*American Journal of Industrial Medicine*

meta-analysis of nonexperimental data have been published in recent years. Table 1-2 lists the topics and the journals of publication for 20 meta-analyses of nonexperimental studies published in 1990–1992.

1.3.2   Decision Analysis

Decision analysis has been used for several decades in business, and it is a part of the core curriculum of many programs leading to the master's degree in business administration. Decision analysis is a derivative of game theory, which was described by von Neumann in the 1920s and applied on a widespread basis in economics by the late 1940s (von Neumann and Morgenstern 1947).

As early as 1959, Ledley and Lusted (1959) described the intellectual basis for the application of decision analysis to medical problems, but it was not until 1967 that a published application of decision analysis to a specific clinical problem appeared. In that year, Henschke and Flehinger (1967) published a paper that used decision analysis to address the question of whether to do radical neck dissection in patients with oral cancer and no palpable neck metastases.

More attention to decision analysis followed publication of articles by Lusted (1971) and Kassirer (1976) that described again the application of decision theory in medicine. The use of decision analysis grew steadily through the 1970s, and by 1981 *Medical Decision Making,* a journal focusing on decision analysis, began publication.

Decision analysis encompasses the use of decision-analytic techniques by clinicians at the bedside as a way to guide the management of individual patients. Recently, the application of decision theory to questions of how to manage individual patients has been less prominent than the use of decision analysis to address policy questions about the use of treatments for groups of patients.

1.3.3 Cost-Effectiveness Analysis

The commonsense principles of cost-benefit and cost-effectiveness analysis have been promoted for centuries (Warner and Luce 1982). Formal use of the techniques in fields other than medicine date to the beginning of the twentieth century. Cost-benefit analysis has been used on a widespread basis in business for at least several decades.

Examples of the systematic analysis of costs in relation to benefits in the medical arena began to appear with some frequency in the mid 1960s. This era has been described as a high point for cost-benefit and cost-effectiveness analysis in the federal health bureaucracy (Warner and Luce 1982). Cost-benefit and cost-effectiveness analysis were particularly prominent in the discussions about treatment of end-stage renal disease that took place in the mid to late 1960s (Warner and Luce 1982). Cost-effectiveness analysis has recently regained prominence in federal discussions of health care reform.

In recent years, cost-benefit analysis has fallen out of favor as a method of addressing health issues, mostly because of the belief that placing monetary values on the outcomes of medical care is impossible or immoral, or both. Use of cost-effectiveness analysis, on the other hand, has been steady.

1.4 LINKAGES OF THE THREE METHODS

The linkage of meta-analysis, decision analysis, and cost-effectiveness analysis to address a single problem is illustrated in two studies of adjuvant chemotherapy in women with early breast cancer, one by the Early Breast Cancer Trialists' Collaborative Group (1988) and the second by Hillner and Smith (1991). After surgical treatment of early breast cancer, a physician must decide whether to recommend treatment with adjuvant chemotherapy—either tamoxifen or cytotoxic chemotherapy. To make this decision, it is important to know how much adjuvant chemotherapy prolongs life and whether it is equally useful in all patients with early breast cancer. The cost of adjuvant chemotherapy needs to be considered in relation to other available therapies for breast cancer.

By 1988, fifty-nine different randomized clinical trials of adjuvant chemotherapy following surgical treatment for breast cancer had been done—28 involving tamoxifen and 31 cytotoxic chemotherapy. These trials involved more than 28,000 women. A meta-analysis by the Early Breast Cancer Trialists' Collaborative Group (1988) showed a highly significant 20% reduction in the annual odds of death within 5 years for women 50 years of age or older who had been treated with tamoxifen, even though none of the 28 trials had individually shown a significant reduction in the odds of dying. Data from these 28 trials are shown in Figure 1-1. There was also a highly significant 22% reduction in the annual odds of death among women under 50 years who had been treated with combination cytotoxic chemotherapy, despite the fact that only 1 of the 31 trials was itself statistically significant. These data are shown in Figure 1-2.

The meta-analysis by the Early Breast Cancer Trialists' Collaborative Group

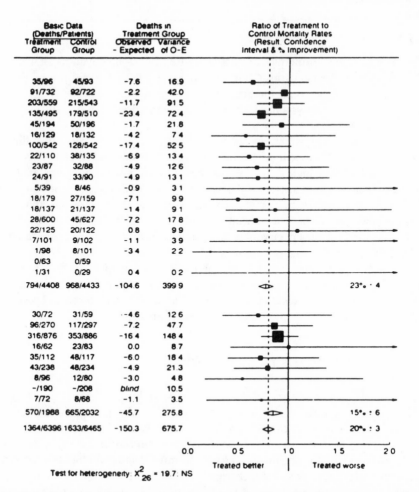

Basic Data (Deaths/Patients)		Deaths in Treatment Group		Ratio of Treatment to Control Mortality Rates (Result, Confidence Interval & % Improvement)
Treatment Group	Control Group	Observed - Expected	Variance of O-E	
35/96	45/93	-7.6	16.9	
91/732	92/722	-2.2	42.0	
203/559	215/543	-11.7	91.5	
135/495	179/510	-23.4	72.4	
45/194	50/196	-1.7	21.8	
16/129	18/132	-4.2	7.4	
100/542	128/542	-17.4	52.5	
22/110	38/135	-6.9	13.4	
23/87	32/88	-4.9	12.6	
24/91	33/90	-4.9	13.1	
5/39	8/46	-0.9	3.1	
18/179	27/159	-7.1	9.9	
18/137	21/137	-1.4	9.1	
28/600	45/627	-7.2	17.8	
22/125	20/122	0.8	9.9	
7/101	9/102	-1.1	3.9	
1/98	8/101	-3.4	2.2	
0/63	0/59			
1/31	0/29	0.4	0.2	
794/4408	968/4433	-104.6	399.9	23% : 4
30/72	31/59	-4.6	12.6	
96/270	117/297	-7.2	47.7	
316/876	353/886	-16.4	148.4	
16/62	23/83	0.0	8.7	
35/112	48/117	-6.0	18.4	
43/238	48/234	-4.9	21.3	
8/96	12/80	-3.0	4.8	
-/190	-/208	blind	10.5	
7/72	8/68	-1.1	3.5	
570/1988	665/2032	-45.7	275.8	15% : 6
1364/6396	1633/6465	-150.3	675.7	20% : 3

Test for heterogeneity: $x^2_{26} = 19.7$; NS

Figure 1-1 For women 50 or more years of age, odds ratio of mortality in women treated with tamoxifen compared with women who were not treated with tamoxifen and 99% confidence interval for this ratio. Each line represents one study. The solid vertical line represents no association with treatment (odds ratio of 1.0). When the line for a study does not cross the solid vertical line, there is a statistically significant association of treatment with a lower risk of death in that study.

The diamonds represent the summary odds ratio for death in treated women and their 95% confidence intervals. When the diamond does not touch the solid vertical line, the summary estimate of the odds ratio of death in treated women is statistically significantly different from 1.0 when based on all of the studies combined. For all studies, there is a highly significant 20% reduction in the odd ratio of death in women 50 years of age or more who were treated with tamoxifen. References to individual studies are as cited in Early Breast Cancer Trialists' Collaborative Group (1988). (Reproduced with permission from Early Breast Cancer Trialists' Collaborative Group, *New England Journal of Medicine,* 1988;319:1685.)

Study Name	Treatment	Basic Data (Deaths/Patients) Treatment Group	Control Group	Deaths in Treatment Group Observed - Expected	Variance of O-E	Ratio of Treatment to Control Mortality Rates (Result, Confidence Interval & % improvement)
(a) Trials of CMF (N.B. includes CMFPr, but not CMF + other cytotoxics)						
INT Milan 7205	CMF	41/85	48/75	-6.3	15.4	
Manchester I *	CMF	8/21	9/24	-0.2	3.8	
Glasgow	CMF	16/47	18/34	-4.6	6.9	
Leiden Mamma	CMF	15/98	23/90	-5.2	8.6	
Danish BCG 77b *	CMF	46/149	44/130	-2.3	20.5	
ECOG EST8177	CMFPr	0/1	1/8	-0.1	0.1	
UK/Asia Collab.	CMF	21/131	20/99	-2.7	9.1	
Ludwig III	CMF	1/2	4/4	-0.4	0.5	
Guy's/Manch. II	CMF	9/63	19/60	-7.0	6.3	
INT Milan 8004	CMF	0/28	5/28	-2.2	1.2	
Danish BCG 82c	CMF	0/4	0/1			
Subtotal (a) CMF		157/635	189/564	-33.3	72.4	37% ± 9
(b) Trials of CMF with extra cytotoxics						
West Midlands	CMFVALeu	48/119	58/119	-6.9	23.0	
Vienna	CMFV	9/32	9/27	-0.8	4.1	
SWOG 7827 A	CMFVPr	-/14	-/13	blind	0.2	
Case Western B	CMFVPr	2/19	0/20	1.0	0.5	
GROCTA Italy	CMF then E	1/38	1/38	0.0	0.5	
Subtotal (b) CMF plus		61/222	68/218	-6.3	28.3	20% ± 17
(c) Trials of regimes without some or all of C, M, F						
Mayo Clinic	CFPr	3/5	1/3	0.8	0.7	
East Berlin	various	22/58	16/28	-1.0	8.9	
DFCI 74083	AC	1/2	1/1	-0.3	0.2	
UK MCCG 003	CVF/CVM	32/61	25/47	2.1	11.4	
Northwick Park	MefV	7/21	8/24	0.0	3.4	
King's CRC I	MeM	35/85	29/62	-1.8	13.6	
West Midlands	LeuMF	19/122	20/131	-0.3	9.2	
Oxford *	MeMF	13/45	19/48	-2.4	7.3	
MD Anderson8026	MV	1/51	6/63	-2.1	1.7	
N Sweden BCG	AC	1/16	7/22	-2.1	1.8	
SE Sweden BCG B	AC	0/0	1/1			
Subtotal (c) other polychem		134/467	133/430	-7.2	56.2	12% ± 13
Total (a + b + c) any polychem		352/1324	391/1202	-46.7	156.9	26% ± 7
(d) Trials of single agents						
Birmingham & WM	C/F	38/77	13/28	1.7	8.7	
NSABP B-05	Mel	24/88	38/81	-7.2	11.8	
Edinburgh II	F	39/59	42/58	-4.3	15.2	
Guy's/Manch. I *	Mel	26/69	34/74	-2.6	13.2	
Dublin	F	5/8	5/10	0.9	2.0	
Danish BCG 77b *	C	39/128	44/130	-2.1	19.0	
Oxford *	Mel	17/44	19/48	-0.3	7.8	
S Swedish BCG	C	17/98	10/90	3.7	6.5	
Subtotal (d) single agents		205/551	205/497	-10.3	84.2	11% ± 10
Total (a + b + c + d) any chem		557/1875	524/1497	-54.6	219.8	22% ± 6

Test for heterogeneity $\chi^2_{28} = 32.2$ NS

0.0   0.5   1.0   1.5   2.0
Treated better | Treated worse

Figure 1-2 For women less than 50 years of age, odds ratio of mortality in women treated with cytotoxic chemotherapy compared with women who did not have chemotherapy and 99% confidence interval for this ratio. Each line represents one study. The symbols are interpreted in the same way as the symbols in Figure 1-1.

For all studies, there is a highly significant 22% reduction in the odds ratio of death in women less than 50 years who were treated with cytotoxic chemotherapy. References to individual studies are as cited in Early Breast Cancer Trialists' Collaborative Group (1988). (Reproduced with permission from Early Breast Cancer Trialists' Collaborative Group, *New England Journal of Medicine*, 1988;319:1686.)

(1988) led the National Cancer Institute to issue a Clinical Alert to all United States physicians. The information showing the benefits of adjuvant chemotherapy for women with early breast cancer was intended to help physicians decide whether to recommend the chemotherapy to their patients with early breast cancer. The analysis strongly suggested that tamoxifen would be useful in women 50 years of age or older with early breast cancer and that cytotoxic chemotherapy would be useful in younger women.

Hillner and Smith (1991) used this meta-analysis (along with data from several of the largest clinical trials) and decision analysis to examine the benefits of adjuvant chemotherapy considering length of life and taking into account other patient and tumor characteristics. Using an assumption that adjuvant chemotherapy reduced the odds of death by 30%, close to the value suggested in the Early Breast Cancer Trialists' Collaborative Group meta-analysis (1988), Hillner and Smith showed that the benefit of adjuvant chemotherapy was highly dependent on the likelihood of having a recurrence of breast cancer, itself a function of age and certain features of the tumor (e.g., size, estrogen receptor status). In a woman with an annual probability of recurrent breast cancer of 1%, as expected in a very small, estrogen receptor positive tumor in a 60-year-old woman, treatment was estimated to increase quality-adjusted life expectancy[4] by only 1 month. In contrast, in a woman with an 8% annual probability of recurrent breast cancer, as expected in a large, estrogen receptor negative tumor in a 45-year-old woman, treatment was estimated to increase quality-adjusted life expectancy by 8 months.

The decision analysis suggested that women at low risk of recurrence have little to gain by adjuvant chemotherapy, whereas high-risk women have much to

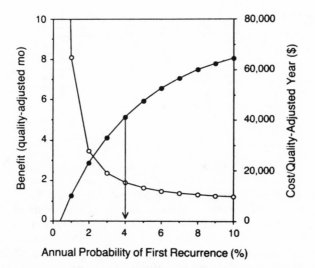

Figure 1-3 In 45-year-old women with node-negative breast cancer, estimated benefit of adjuvant chemotherapy compared with no treatment, according to the annual probability of recurrence of breast cancer. The left vertical axis shows the benefit in terms of quality-adjusted life months; the line with solid circles shows the benefit by this measure. The right vertical axis shows the benefit in terms of cost per quality-adjusted life year; the line with the open circles shows the benefit by this measure. The greater the probability of recurrence the greater the benefit in terms of gain in quality-adjusted life months and the less the cost per quality-adjusted life year gained. (Reproduced with permission from Hillner and Smith, *New England Journal of Medicine*, 1991;324:164.)

gain. By expressing the benefit in terms of life expectancy, the decision analysis translates the results of the meta-analysis into units that are more intuitively accessible to patients and to physicians. The delineation of underlying risk of breast cancer recurrence as an important determinant of the benefit of adjuvant chemotherapy allows some women to be spared the inconvenience and discomfort associated with chemotherapy. Thus, recommendations by physicians about adjuvant chemotherapy can be tailored more carefully to the individual patient.

Hillner and Smith (1991) also used cost-effectiveness analysis to estimate the cost of adjuvant chemotherapy overall and for the subgroups defined by their risk of breast cancer recurrence. As shown in Figure 1-3, the estimated cost of adjuvant chemotherapy for a woman with a 1% annual probability of recurrent breast cancer was about $65,000 per quality-adjusted life year. For a woman with an 8% annual probability of recurrence, the estimated cost of adjuvant chemotherapy was only $9,000 per quality-adjusted life year.

The cost-effectiveness analysis allows choices between adjuvant chemotherapy and competing therapies to be made. It also allows for comparison of adjuvant chemotherapy with other life-enhancing interventions that might be offered to the same woman.

Taken together, the meta-analysis, the decision analysis, and the cost-effectiveness analysis for this topic provide a rational framework for counseling patients with early breast cancer, setting clinical policies about use of adjuvant chemotherapy, and aiding in policy decisions about funding of these therapies. The integration of the three research methods illustrates the power of the three techniques as clinical and policy tools and foreshadows their future use.

1.5 Organization of the Book

Chapter 2 gives an overview of each method as applied in its simplest form. The subsequent three chapters describe three elements—planning, information retrieval, and data collection—which inhere to common principles. Chapters 6, 7, and 8 present the advanced features of meta-analysis. Chapters 9, 10, and 11 present the advanced features of decision analysis. Chapter 12 shows how to do cost-effectiveness analysis based on the results of a decision analysis. Chapter 13 describes sensitivity analysis as it applies to all three methods. Chapter 14 suggests ways to present effectively the results of studies using each method. Chapter 15 describes the most critical limitations of the methods and identifies the situations in which the methods are the most and least useful.

NOTES

1. The term "clinical decision analysis" is used to refer to decision analysis as applied to management of individual patients at the bedside. This book addresses decision analysis as it applies to formulation of policy decisions for groups of patients. The reader interested in clinical decision analysis is referred to Sox et al. (1988).

2. Measures of value are referred to broadly as measures of "utility." The term "utility analysis" is often used to refer to analysis that uses measures of value other than lives saved

or life expectancy. The most common measure of value used in utility analysis is the quality-adjusted life year.

3. Sisk and Riegelman also did a cost-effectiveness analysis of pneumococcal vaccine from the societal perspective, but only the analysis taking the Medicare perspective is discussed in this chapter.

4. Quality adjustment of life expectancy attempts to take into account not only the length of life but also the quality of life during the period of extended life. It is an acknowledgment that there are fates worse than death.

2
Overview of the Methods

Doing a study that involves meta-analysis, decision analysis, or cost-effectiveness analysis is complex for most real life applications, reflecting the complexity of the problems that the methods are used to address. However, each method involves a limited number of discrete, fairly simple steps. The three sections of this chapter give an overview of meta-analysis, decision analysis, and cost-effectiveness analysis and describe the steps in applying the methods in their simplest form. In later chapters, advanced issues in application of the three methods are discussed in depth.

2.1 META-ANALYSIS

2.1.1 Overall Goals, Main Uses, and Description of Steps

The overall goal of meta-analysis is to combine the results of previous studies to arrive at summary conclusions about a body of research. It is most useful in summarizing prior research when individual studies are small and they are individually too small to yield a valid conclusion.

EXAMPLE: In 1982, use of thrombolytic agents after acute myocardial infarction was controversial. At that time, eight randomized clinical trials examining the effect of a loading dose of at least 250,000 international units of intravenous streptokinase on mortality after an acute myocardial infarction had been done (Stampfer et al. 1982). As shown in Table 2-1, two of

Table 2-1 Results of randomized trials of effect on mortality of intravenous streptokinase following acute myocardial infarction published before 1982

	N Deaths/Total		Mortality (%)		Estimated Relative Risk (95% Confidence
Reference	Treated	Control	Treated	Control	Interval)
Avery et al. (1969)	20/83	15/84	24.1	17.9	1.35 (0.74–2.45)
European Working Party (1971)	69/373	94/357	18.5	26.3	0.70 (0.53–0.92)[a]
Heikinheimo et al. (1971)	22/219	17/207	10.0	8.2	1.22 (0.67–2.24)
Dioguardia et al. (1971)	19/164	18/157	11.6	11.5	1.01 (0.55–1.85)
Breddin et al. (1973)	13/102	29/104	12.7	27.9	0.46 (0.26–0.81)
Bett et al. (1973)	21/264	23/253	8.0	9.1	0.88 (0.50–1.54)
Aber et al. (1976)	43/302	44/293	14.2	15.0	0.95 (0.64–1.40)
European Cooperative Study Group for Streptokinase in Acute Myocardial Infarction (1979)	18/156	30/159	11.5	18.9	0.61 (0.36–1.04)
			Summary relative risk		0.80 (0.68–0.95)

[1]$p < 0.01$.

Source: Stampfer et al. (1982); table references cited there.

the trials found a higher risk of mortality in treated patients, five found a lower risk, and one found essentially identical mortality in the treated and the control patients. The trials were all fairly small, and the difference in mortality between treated and control patients was statistically significant in only one trial. These studies were interpreted as inconclusive about the benefit of intravenous streptokinase.

In a meta-analysis based on these trials, Stampfer et al. (1982) estimated the relative risk of mortality in patients treated with intravenous streptokinase to be 0.80 with 95% confidence limits of 0.68 and 0.95. A subsequent studies of intravenous streptokinase after acute myocardial infarction involving thousands of patients (GISSI 1986) confirmed the conclusion based on the meta-analysis of early studies—that intravenous treatment with streptokinase reduces mortality following acute myocardial infarction.

Meta-analysis has been applied most often to combine the results of randomized trials. However, there are many topics for which randomized trials are impossible. For example, smoking and alcohol use cannot be assigned at random. Meta-analysis of nonexperimental studies is also common. For nonexperimental studies, the method is also most useful when there are many studies with low statistical power.

EXAMPLE: The effect of exposure to environmental tobacco smoke on lung cancer risk is a topic of considerable public health importance. By 1991, there were 19 case-control studies of lung cancer in which information on exposure to environmental tobacco smoke was available in addition to

Table 2-2 For 19 case-control studies, number of cases of lung cancer in women who did not actively smoke cigarettes and estimated relative risk of lung cancer in relation to exposure to environmental tobacco smoke

Reference	Number of Cases	Estimated Relative Risk (95% Confidence Interval)
Akiba, Kato, Blot (1986)	94	1.52 (0.88–2.63)
Brownson et al. (1987)	19	1.52 (0.39–5.99)
Buffler et al. (1984)	41	0.81 (0.34–1.90)
Chan et al. (1979)	84	0.75 (0.43–1.30)
Correa et al. (1983)	22	2.07 (0.82–5.25)
Gao et al. (1978)	246	1.19 (0.82–1.73)
Garfinkel, Auerbach, Joubert (1985)	134	1.31 (0.87–1.98)
Geng, Liang, Zhang (1988)	54	2.16 (1.08–4.29)
Humble, Samet, Pathak (1987)	20	2.34 (0.81–6.75)
Inoue, Hirayama (1988)	22	2.55 (0.74–8.78)
Kabat, Wynder (1984)	24	0.79 (0.25–2.45)
Koo et al. (1987)	86	1.55 (0.90–2.67)
Lam et al. (1987)	199	1.65 (1.16–2.35)
Lam (1985)	60	2.01 (1.09–3.71)
Lee, Chamberlain, Alderson (1986)	32	1.03 (0.41–2.55)
Pershagen, Hrubec, Svensson (1987)	67	1.28 (0.76–2.15)
Svensson, Pershagen, Klominek (1988)	34	1.26 (0.57–2.82)
Trichopoulos, Kalandidi, Sparros (1983)	62	2.13 (1.19–3.83)
Wu et al. (1985)	28	1.41 (0.54–3.67)
Summary relative risk		1.42 (1.24–1.63)

Source: Environmental Protection Agency (1990); table references cited there.

information on active smoking. The strong effect of active cigarette smoking on lung cancer made it impossible to separate the effect of active smoking from other sources of environmental tobacco smoke in smokers, and the question of an effect of environmental tobacco smoke on lung cancer risk had to be addressed using lung cancer cases in nonsmokers. The possibility of confounding due to occupational exposure made it desirable to restrict analysis to women. As shown in Table 2-2, the number of cases of lung cancer in women who never smoked cigarettes was small in each of the 19 case-control studies. Fifteen studies found an estimated relative risk greater than 1; three found an estimated relative risk less than 1; one study found essentially no increase or decrease in estimated relative risk.

In a meta-analysis of these data by scientists at the Environmental Protection Agency (USEPA 1990), the relative risk of lung cancer in women exposed to environmental tobacco smoke was estimated to be 1.42 with 95% confidence limits of 1.24 and 1.63. This information was the basis for a decision by an advisory committee of the Environmental Protection Agency to designate environmental tobacco smoke as a carcinogen.

There are four steps in a meta-analysis. First, studies with relevant data are identified. Second, eligibility criteria for inclusion and exclusion of the studies are defined. Third, data are abstracted. Fourth, the abstracted data are analyzed statistically.

2.1.2 Identifying Studies for the Meta-Analysis

A critical feature of the proper application of the method of meta-analysis is development of systematic, explicit procedures for identifying studies with relevant data. The systematic, explicit nature of the procedures for study identification distinguishes meta-analysis from qualitative literature review. In being systematic, the procedures reduce bias. In being explicit, the procedures help to ensure reproducibility. No matter how sophisticated the statistical techniques used to aggregate data from studies, a review does not qualify as meta-analysis unless the procedures to identify studies are both systematic and explicit.

Identification of published studies usually begins with a search of personal reference files and is followed by a computerized search of MEDLINE and of other computerized literature databases. The title and abstract of studies identified in the computerized search are scanned to exclude any that are clearly irrelevant. The full text of the remaining articles is retrieved, and each paper is read to determine whether it contains information on the topic of interest. The reference lists of articles with information on the topic of interest are reviewed to identify citations to other studies of the same topic, and publications that were not identified in the computerized literature search are retrieved and reviewed for presence of relevant information. Reference lists of review articles are also reviewed to check for completeness of the assembled list of relevant publications. In many cases, the list of studies identified by computer literature search and reference checks is submitted for review to a knowledgeable expert, who is asked to identify studies of the topic that have not been included on the list.

> *EXAMPLE:* Dupont and Page (1991) set out to identify publications that presented information on menopausal estrogen replacement therapy and breast cancer. They used MEDLINE to identify 556 articles indexed with a MeSH heading of "breast neoplasms" and either "estrogens" or "estrogens, synthetic" and were also classified under the MeSH category "occurrence," "etiology," "epidemiology," or "chemically induced" and in the MeSH category "human" and "female." Thirty-five publications identified in the MEDLINE search provided an estimate of breast cancer risk in women who took estrogen replacement therapy. The reference lists of these 35 publications and those in a review article led to identification of 15 more publications with information on breast cancer risk in women using estrogen replacement therapy.

Details of procedures for identifying studies for a meta-analysis are discussed in Chapter 4.

2.1.3 Defining Eligibility Criteria for the Meta-Analysis

After studies with relevant information have been identified, the next step in the meta-analysis is to define eligibility criteria for the meta-analysis. Just as not all people are eligible for a randomized trial, not all studies can or should be included in the meta-analysis. For example, nonexperimental studies usually are not eligible for a meta-analysis of randomized trials; studies of stroke should not be eligible for a meta-analysis of coronary heart disease; and studies of a nontherapeutic dose of a drug should not be included in a meta-analysis of the efficacy of the drug.

EXAMPLE: The meta-analysis of intravenous streptokinase and mortality after acute myocardial infarction by Stampfer et al. (1982) excluded four studies because a careful reading of the methods sections for these four studies showed that allocation to the treatment and control groups was not strictly random.

The goals of defining eligibility criteria are to ensure reproducibility of the meta-analysis and to minimize bias in selection of studies for the meta-analysis. Another analyst faced with the same body of literature applying the same eligibility criteria should choose the same set of studies. The studies chosen for the meta-analysis should be unbiased with respect to their results and their conclusions.

EXAMPLE: Early studies of intravenous streptokinase used loading doses that ranged from several thousand to over 1 million international units. If dose is an important determinant of the effect of streptokinase on mortality, inclusion of studies with very low doses might bias the meta-analysis toward finding no effect of the drug. Recognizing this, Stampfer et al. (1982) restricted their meta-analysis to studies that used a loading dose at least 250,000 international units of streptokinase.

Additional detail on defining the eligibility criteria for a meta-analysis appears in Chapter 6.

2.1.4 Abstracting Data

In a meta-analysis, there are usually two levels of data abstraction. First, data that document whether or not identified studies are eligible for the meta-analysis study need to be abstracted for all of the studies identified. Next, for all eligible studies, data on the relevant outcomes of the study and the characteristics of the study, such as number of patients, are abstracted. The procedures for abstracting data in a meta-analysis should be similar to procedures to abstract data from a medical record or other administrative document. That is, data should be abstracted onto structured forms that have been pretested, and an explicit plan to ensure reliability of abstraction should be in place.

Greater detail on development of forms and on ensuring reliability of data abstraction appears in Chapter 5.

2.1.5 Analyzing the Data Statistically

The last step in a meta-analysis is to analyze the data statistically. This step most often includes combining the data to arrive at a summary estimate of the effect size, a measure of its variance and its 95% confidence interval, and a test for homogeneity of effect size. It may include examination of the reasons for heterogeneity. Models of dose-response may be developed, and the predictors of effect size may be explored in multivariate models. Chapters 7 and 8 describe the statistical methods to derive summary estimates of effects based on both fixed-effects and random-effects models and some of the modeling techniques as they are applied in meta-analysis.

2.2 DECISION ANALYSIS

2.2.1 Overall Goals, Main Uses, and Description of Steps

Decision analysis is a systematic quantitative approach for assessing the relative value of one or more different decision options. Historically, it has been described as a method to help clinicians make decisions on how to manage individual patients (Weinstein and Fineberg 1980; Sox et al. 1988). It is increasingly used to help develop policies about the management of groups of patients by providing information on which of two or more strategies for approaching a medical problem has the "best" outcome or the most value. Decision analysis is also the first step in a cost-effectiveness analysis, and it is used increasingly for this purpose.

Decision analysis is useful when the clinical or policy decision is complex and information is uncertain. The method is particularly useful in examining issues when at least some of the consequences of the decision are distant in time from the decision.

EXAMPLE: Gallstones are often detected in persons who have no symptoms of gallstone disease. In such persons, a decision must be made on whether to do a "prophylactic" cholecystectomy or to wait until symptoms develop to operate. Ransohoff et al. (1983) did a decision analysis to compare the effect on life expectancy of prophylactic cholecystectomy and expectant waiting.

If a person has a prophylactic cholecystectomy, an immediate consequence is the possibility of operative death. If a person does not have a prophylactic cholecystectomy, possible consequences are death from other causes before the gallstones cause symptoms or development of pain or another complication of biliary disease before death from another cause, events that would require a cholecystectomy. Operative mortality after cholecystectomy is influenced by the age of the patient and by the presence during the operation of complications of gallstone disease, such as acute cholecystitis. The decision about whether to do a prophylactic cholecystectomy or to wait is complex at least in part because the consequences of waiting are far removed in time from the decision about whether or not to operate.

The analysis by Ransohoff et al. (1983) showed that a decision to do prophylactic cholecystectomy would result in an average loss of 4 days of life for a 30-year-old man and 18 days for a 50-year-old man. The analysis supports a decision to forgo prophylactic cholecystectomy.

There are five steps in a decision analysis (Weinstein and Fineberg 1980). First, the problem is identified and bounded. Second, the problem is structured, a process that usually includes construction of a decision tree. Third, information necessary to fill in the decision tree is gathered. Fourth, the decision tree is analyzed. Last, a sensitivity analysis is done.

2.2.2 Identifying and Bounding the Problem

The first step in a decision analysis is to identify and bound the problem (Weinstein and Fineberg 1980). Problem identification consists of stating the main issue concisely. Identifying and bounding the problem consists of breaking the problem down into its components. The first component of a problem is always identification of the alternative courses of action.

> *EXAMPLE:*   Fifteen new cases of measles are reported in a small urban area. This is the first report of measles in the area in several years. All of the cases are in children age 8 through 15 who previously received only one measles vaccination. This schedule was recommended at the time these children were infants, but it is now known not to confer complete and lifelong immunity to measles in all persons who are vaccinated. The problem is deciding whether to recommend that children who were vaccinated only once be revaccinated. The first component of the problem is identification of the alternative courses of action. One course of action is to recommend revaccination for all children 8 through 15; the alternative is not to recommend revaccination.

Other components of the problem are then identified. These are usually events that follow the first course of action and its alternative. The final component of the decision problem is identification of the outcome.

> *EXAMPLE:*   The relevant event that follows revaccinating or not revaccinating children is exposure to an infectious case of measles. Upon exposure to an infectious case, children either contract or do not contract measles. If they contract measles, the outcome of interest (for the purpose of this example) is death from measles.

2.2.3 Structuring the Problem

To structure the problem in a decision analysis, a decision tree is constructed. The decision tree depicts graphically the components of the decision problems and relates actions to consequences (Schwartz et al. 1973).

The building of a decision tree is guided by a number of conventions. Thus, by convention, a decision tree is built from left to right. When time is an issue, earlier events and choices are depicted on the left and later ones on the right.

A decision tree consists of nodes, branches, and outcomes. There are two kinds of nodes—decision nodes and chance nodes. Decision nodes are, by convention,

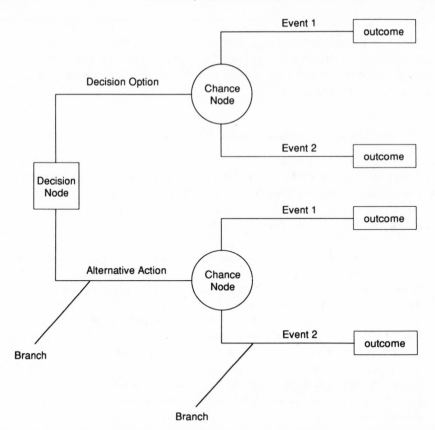

Figure 2-1 Hypothetical decision tree. The decision node is represented with a square. Chance nodes are represented with circles. Outcomes are represented with rectangles. Branches are drawn at right angles to the decision and chance nodes.

depicted as squares. Chance nodes are depicted as circles. Outcomes are depicted as large rectangles. Branches are conventionally drawn at right angles to nodes; they connect nodes with nodes and nodes with outcome.

EXAMPLE: Figure 2-1 is the skeleton of a decision tree with nodes, branches, and outcomes labeled.

Decision nodes identify points where there are alternative actions that are under the control of the decision maker. In the simplest problem, the decision node describes the problem.

EXAMPLE: Figure 2-2 shows the beginning of a decision tree for the problem of whether or not to recommend measles revaccination of children 8 to 15. The square decision node at the left of the diagram represents the decision; the alternative courses of action—to recommend revaccination or not to recommend revaccination—are labeled on the horizontal portions of the branches.

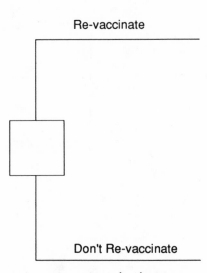

Re-vaccinate

Don't Re-vaccinate

Figure 2-2 The first step in construction of a decision tree for the measles revaccination problem. The decision node is drawn as a square. The two alternatives—revaccinate and do not revaccinate—are represented as branches.

Chance nodes identify points where one or more of several possible events that are beyond the control of the decision maker may occur. Chance nodes for the same events should line up horizontally in the decision tree.

Probabilities are associated with the events depicted at chance nodes. At any given chance node, the sum of the probabilities of the events must be equal to 1. That is, the chance node defines events that are mutually exclusive and jointly exhaustive.

EXAMPLE: Figure 2-3 is a decision tree for the measles revaccination problem. The circular chance nodes identify the first event that follows the decision to revaccinate—either children are exposed to measles or they are not exposed to measles. This event is out of the control of the decision maker. The sum of the probabilities of being exposed or not being exposed to measles is 1.

Figure 2-4 is a decision tree for the measles problem with circular chance nodes to also identify events that follow the exposure to measles. Children exposed to measles either get measles or they do not get measles. Again, this is an event that is out of the control of the decision maker, and it is depicted by a chance node. The sum of the probabilities of getting or not getting measles is 1.

In the decision tree, outcomes are the consequences of the final events depicted in the tree. Outcomes may include life or death; disability or health; or any of a variety of other risks or benefits of the treatment.

EXAMPLE: The rectangular boxes in Figure 2-5 identify the outcome of getting and of not getting measles. For this example, the outcomes of interest are death or nondeath from measles.

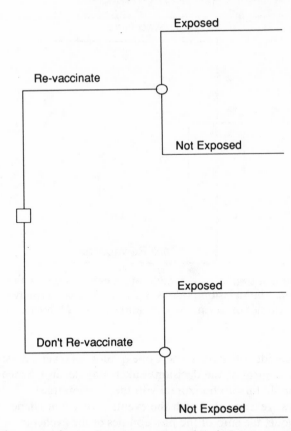

Figure 2-3　The second step in construction of a decision tree for the measles revaccination problem. Chance nodes that represent the likelihood of being exposed to measles are drawn.

Most current decision analyses do not focus simply on the comparison of decision options in terms of their effect on life and death. They focus on the amount of extension in life and on measures of the quality of life. This focus recognizes the use of medical care to do things other than prevent death. Moreover, everyone dies, and analyses of medical interventions can reasonably expect only to delay death, not to prevent it. Outcome measures used in most current decision analyses are called *utilities*. A utility is a measure of value to society or to an individual. Chapter 11 is devoted to a description of the concept of utilities, the measurement of utilities, and the incorporation of utility measures into a decision analysis.

2.2.4　Gathering Information to Fill in the Decision Tree

The next step in the decision analysis is to gather information on the probabilities of each chance event. Information gathering for decision analysis almost always uses one or more of the following: literature review, including meta-analysis; primary data collection; and consultation with experts. After the information on the probabilities and the outcome is obtained, it is recorded on the decision tree.

EXAMPLE: In the context of an epidemic of measles in an inner-city population, experts estimate that 20 out of every 100 children age 8 through 15 will come in contact with an infectious case of measles each year. Literature review reveals that the probability of getting measles if exposed to an infectious case is 0.33 in a child who has had only one measles vaccination and 0.05 in a child who is revaccinated (Mast et al. 1990). The probability of getting measles in children who are not exposed to measles is, of course, zero. During the current epidemic, the probability of dying from measles if a child gets measles is 23 per 10,000 cases, or 0.0023 (Centers for Disease Control 1990). It is assumed that the probability of dying from measles in children who don't get measles is zero. Figure 2-6 shows the decision tree on which these probability estimates are shown.

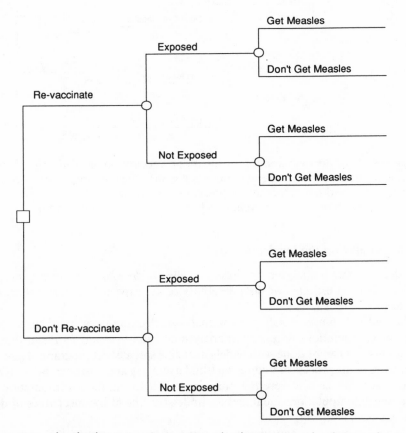

Figure 2-4   The third step in construction of a decision tree for the measles revaccination problem. For each branch of the exposed/not exposed dichotomy represented in the prior step, the chance of getting or not getting measles is represented with a chance node.

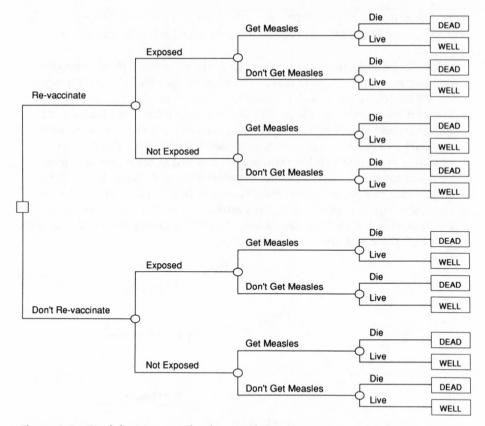

Figure 2-5 Final decision tree for the measles revaccination problem. The chances of dying or remaining alive after getting or not getting measles are represented with chance nodes, and the outcome is represented as a rectangle. In this problem, the outcome is death or remaining well, which is the same as the final event.

2.2.5   Analyzing the Decision Tree

The decision tree is analyzed by a process called *folding back* and *averaging*. The final result is an estimate of the probability of the expected outcome of each of the decision alternatives.

Specialized computer software for analyzing decision trees is available. However, the computations necessary to analyze decision trees are simple arithmetic operations that can be done with widely available spreadsheet programs. Here, the mechanics of the process of folding back and averaging are illustrated by analyzing the decision tree as if it were two spreadsheets. Showing these computations as spreadsheet computations facilitates an understanding of the mechanics of decision analysis.

The decision tree is considered to consist of spreadsheets, one for each of the decision alternatives. The number of rows in each spreadsheet is equal to the number of outcome boxes in the decision tree. The spreadsheet has a column for each probability estimate and a column for the estimated probability of the outcome.

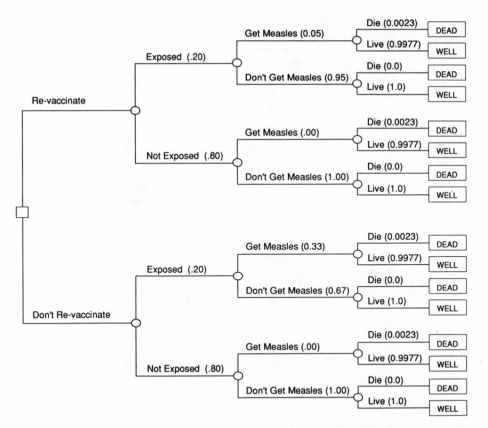

Figure 2-6 Measles decision tree showing all of the probabilities used in the analysis.

EXAMPLE: Table 2-3 recasts the measles problem as two blank spreadsheets—one for the revaccination decision option and one for the no-revaccination option. For each spreadsheet, there are eight rows, because there are eight outcome boxes in the decision tree. There are three columns, one for each probability estimate and one for the outcome.

After the spreadsheets are set up, the probabilities are filled in.

EXAMPLE: Table 2-4 shows the spreadsheet for the revaccination arm of the decision tree with the relevant probabilities in their proper columns.

The next step is to carry out the process of folding back and averaging. For each row, all of the probabilities in the row are multiplied together. This is folding back the decision tree. The products of the rows that represent the same outcome (die or don't die) are summed for each decision option. This is averaging. The sum of the products is the expected value of that outcome for the specified decision option.

Table 2-3 Measles decision analysis as a spreadsheet

Revaccination		
Probability of Exposure	Probability of Getting Measles	Probability of Outcome
		die
		don't die
		die
		don't die
		die
		don't die
		die
		don't die

No Revaccination		
Probability of Exposure	Probability of Getting Measles	Probability of Outcome
		die
		don't die
		die
		don't die
		die
		don't die
		die
		don't die

EXAMPLE: Table 2–5 shows the measles problem with a column labeled "product" for each row. The number in the column labeled "product" is the product of the probabilities for the corresponding row. For example, the value in the first row of the column labeled "product" is 0.000023, which is

$$0.2 \times 0.05 \times 0.0023$$

The expected probability of death from measles in the example is equal to the sum of the values in the product column for the rows that correspond to the outcome "die." For the revaccination option, this is

$$0.000023 + 0.000000 + 0.000000 + 0.000000 = 0.000023$$

For the no-revaccination option, it is

$$0.000152 + 0.000000 + 0.000000 + 0.000000 = 0.000152$$

The final step is to compare the two strategies by subtracting the results of the preceding calculations for the revaccination arm from the result for the alternative arm.

EXAMPLE: The difference in the expected probability of death from measles between a strategy of revaccination and a strategy of no-revaccination is

$$0.000152 - 0.000023 = 0.000129$$

This is interpreted to mean that 12.9 deaths from measles are prevented per 100,000 children revaccinated.

2.2.6 Sensitivity Analysis

Analysis of a decision tree virtually always includes sensitivity analysis. Sensitivity analysis is described in detail in Chapter 13. The overall goal of sensitivity analysis is to compare the stability of the conclusion of the analysis to assumptions made in the analysis. Sensitivity analysis also may identify crucial areas of information deficiency and may guide further research.

Assumptions about the probabilities used in the analysis are among the most important assumptions made in the analysis. A sensitivity analysis varying these

Table 2-4 Measles decision analysis as a spreadsheet

Revaccination			
Probability of Exposure	Probability of Getting Measles	Probability of Outcome	
0.2	0.05	0.0023	die
0.2	0.05	0.9977	don't die
0.2	0.95	0.0000	die
0.2	0.95	1.0000	don't die
0.8	0	0.0023	die
0.8	0	0.9977	don't die
0.8	1	0.0000	die
0.8	1	1.0000	don't die

No Revaccination			
Probability of Exposure	Probability of Getting Measles	Probability of Outcome	
0.2	0.33	0.0023	die
0.2	0.33	0.9977	don't die
0.2	0.67	0.0000	die
0.2	0.67	1.0000	don't die
0.8	0	0.0023	die
0.8	0	0.9977	don't die
0.8	1	0.0000	die
0.8	1	1.0000	don't die

Table 2-5   Measles decision analysis as a spreadsheet

		Revaccination		
Product	Probability of Exposure	Probability of Getting Measles	Probability of Outcome	
0.000023	0.2	0.05	0.0023	die
0.009977	0.2	0.05	0.9977	don't die
0.000000	0.2	0.95	0.0000	die
0.190000	0.2	0.95	1.0000	don't die
0.000000	0.8	0	0.0023	die
0.000000	0.8	0	0.9977	don't die
0.000000	0.8	1	0.0000	die
0.800000	0.8	1	1.0000	don't die

Sum for deaths
0.000023

		No Revaccination		
Product	Probability of Exposure	Probability of Getting Measles	Probability of Outcome	
0.000152	0.2	0.33	0.0023	die
0.065848	0.2	0.33	0.9977	don't die
0.000000	0.2	0.67	0.0000	die
0.134000	0.2	0.67	1.0000	don't die
0.000000	0.8	0	0.0023	die
0.000000	0.8	0	0.9977	don't die
0.000000	0.8	1	0.0000	die
0.800000	0.8	1	1.0000	don't die

Sum for death
0.00152

Difference between revaccination and no revaccination
Death   0.000129

Difference expressed as events per 100,000
Death   12.9

probabilities one at a time while holding all of the other variables in the analysis constant is almost always done.

EXAMPLE: The probability of being exposed to an infectious case of measles varies according to the area of the county. It is 1 per 100 in the suburban areas, whereas it is 45 per 100 in one inner-city area where an epidemic is in progress. The results of a sensitivity analysis varying the probability of exposure to an infectious case of measles between 0.01 and 0.45 is shown in Table 2–6. The number of deaths from measles prevented per 100,000 children revaccinated is highly dependent on the assumption about the probability of being exposed to an infectious case. Revaccination is esti-

Table 2-6 Results of sensitivity
analysis varying probability of
exposure to measles

Assumed Probability of Exposure	Net Number of Lives Saved per 100,000 Children Revaccinated
0.01	0.6
0.05	3.2
0.10	6.4
0.15	9.7
0.20[a]	12.9
0.25	16.1
0.30	19.3
0.35	22.5
0.40	25.8
0.45	29.0

[a]Probability in baseline analysis.

mated to prevent less than 1 death from measles per 100,000 children revaccinated in the low-risk area and 29 in the highest risk area.

2.3 COST-EFFECTIVENESS ANALYSIS

2.3.1 Overall Goals, Main Uses, and Description of Steps

Cost-effectiveness analysis compares the outcome of decision options in terms of their monetary cost per unit of effectiveness. It is used to try to set priorities for the allocation of resources and to decide among one or more treatments or interventions. Cost-effectiveness analysis is most useful when it compares alternative treatments for the same condition.

EXAMPLE: End-stage renal disease can be treated in four ways: home dialysis, in-center dialysis, transplant from a living related donor, and cadaver transplant. All four approaches prolong life, but they differ in their effect on life expectancy and in their cost. Garner and Dardis (1987) did a cost-effectiveness analysis comparing the costs of the four treatments. Considering patients treated for 20 years, home dialysis and transplantation were more cost effective than in-center dialysis. In this example, the question posed is not whether to treat end-stage renal disease, but *how* to treat it.

The first four steps in a cost-effectiveness analysis are the same as in a decision analysis. The problem is identified and bounded, a decision tree is constructed, information to fill in the decision tree is gathered, and the decision tree is analyzed to determine the outcome of the decision options.

Cost-effectiveness analysis involves the following additional steps. First, the perspective of the analysis is defined and, based on consideration of the perspective, information on relevant costs is gathered. Second, the information on cost of each decision option is analyzed to determine the net cost of each of the decision options. The net cost of each decision option per unit of the outcome measure is computed for each of the decision options. Third, the decision options are compared in relation to net cost per unit of outcome. Last, a sensitivity analysis is done.

2.3.2   Defining the Perspective

Costs are seen differently from different points of view. For example, the cost of hospitalization from the perspective of an insurance company is the amount of money that the company pays the hospital for that illness under the coverage plan for the individual who is hospitalized. The cost from the perspective of the hospital is the true cost of providing the service, which includes the labor costs, the costs of the building in which the services are provided, and other overhead costs. It is important to state explicitly the perspective of a cost-effectiveness analysis, since the perspective determines which costs should be included in the analysis and what economic outcomes are considered as benefits. Usual perspectives in cost-effectiveness analysis are the societal perspective and the program perspective.

EXAMPLE: The decision about whether to revaccinate children against measles could be made based on considerations of cost. A county health department might decide to undertake a cost-effectiveness analysis of revaccinating versus not revaccinating taking the program perspective, since the question addressed is how many deaths from measles an investment in revaccination might prevent. Taking the program perspective, the costs that will be considered are the costs that are born directly by the program.

2.3.3   Obtaining Cost Data

Obtaining data on costs is one of the most difficult and important parts of a cost-effectiveness analysis. To obtain proper cost data requires an understanding of the meaning of the term "cost" in economics, an appreciation of the difference between costs and charges, and an understanding of the effect of the perspective of the analysis on the types of cost data to be included. In addition, data on cost must be collected either primarily (i.e., as a special study) or from secondary sources. These topics are covered in more detail in Chapter 12.

Cost data are generally gathered either from administrative sources, such as insurance companies, or by carrying out special studies that involve retrieval of billings of samples of patients.

EXAMPLE: By reviewing the literature, the county determines that the cost of measles vaccine in bulk quantities is $4.44 per dose (Mast et al. 1990). An expert estimates that it will take 15 minutes for a nurse to vaccinate each child in the school setting. Based on a school nurse's salary of

$36,000 per year and considering an overhead cost of $2.00 per child revaccinated, it is estimated that the total cost of revaccination is $9.44 per child.

2.3.4 Analyzing the Cost Data and Using the Decision Analysis

In the simplest case, the decision analysis proceeds as described in the section on decision analysis, yielding an estimate of the net benefit of one decision option compared with the other. In a cost-effectiveness analysis, the net cost of each of the decision options is calculated. The cost of one decision option in relation to the other is computed by subtraction, and the net cost per unit of net outcome is computed by division.

EXAMPLE: Based on the estimate that it will cost $9.44 to revaccinate each child, the cost to the county of a program revaccinating 100,000 children is $944,000; the cost to the county of not revaccinating 100,000 children is $0. The net cost of a program of revaccination compared with no program of revaccination is $944,000. The net number of deaths from measles prevented per 100,000 children revaccinated, estimated in the decision analysis, is 12.9. Thus, compared with a strategy of no-revaccination, a school-based program of measles revaccination costs $73,178 per death prevented, or ($944,000 − $0)/12.9.

When costs and monetary benefits of a treatment are spread over time, it is necessary to discount and to consider inflation. These issues are discussed in detail in Chapter 12.

2.3.5 Sensitivity Analysis

As in decision analysis, sensitivity analysis is almost always done in a cost-effectiveness analysis. It has the same goal—to assess the effect of the various assump-

Table 2-7 Results of sensitivity analysis varying probability of exposure to measles

Assumed Probability of Exposure	Cost per Life Saved
0.01	$1,573,333
0.05	295,000
0.10	147,500
0.15	97,320
0.20[a]	73,178
0.25	58,634
0.30	48,912
0.35	41,956
0.40	36,589
0.45	32,552

[a]Probability in baseline analysis.

tions made in the analysis on the conclusion. Just as in decision analysis, assumptions about the values of the probabilities used in the analysis are usually done.

EXAMPLE: The sensitivity analysis from the decision analysis in which the probability of being exposed to measles was varied is used to do a sensitivity analysis for the cost-effectiveness analysis. In this sensitivity analysis, the cost of the program per 100,000 children, $944,000, is assumed to be fixed. The number of deaths prevented by a program, compared with no program, was estimated in Section 2.2.4 for several estimates of the probability of being exposed to measles. The net cost of the program compared with no program is divided by each of these estimates to estimate the cost per death prevented for several estimates of the probability of exposure to measles. These results are shown in Table 2-7. The estimated cost per death prevented is highly variable, depending on the assumption about the probability of exposure to measles. In low-risk areas, the analysis shows that the cost of the program is over $1.5 million per death prevented; in the highest risk area, it is $32,552 per death prevented.

3

Planning the Study

Studies that involve synthesis seem seductively simple to do. The impression of simplicity has several consequences. First, data may be collected without a formal plan and analyzed in an ad hoc fashion. Second, resource needs for the study may be underestimated. Last, administrative obstacles may not be anticipated. In reality, it is as important to carefully plan a study that involves synthesis as it is to carefully plan a clinical trial, a cross-sectional survey, and a case-control or cohort study. Documentation of all aspects of study design and conduct is a crucial and often overlooked step in carrying out studies that involve meta-analysis, decision analysis, and cost-effectiveness analysis.

The four steps common to the planning all three types of synthetic studies are definition of the problem; development of a protocol; acquisition of resources; and procurement of administrative approvals. The four sections of this chapter describe these four steps.

3.1 DEFINING THE PROBLEM

The first step in planning the study is to define the problem. The problem definition is a general statement of the broad issue that the study addresses.

EXAMPLE: Chapter 1 described a meta-analysis of antiplatelet drugs to prevent stroke in patients with transient ischemic attacks, a decision analysis on treatment of HIV-infected intravenous drug users to prevent tuberculosis, and a cost-effectiveness analysis on pneumococcal vaccination in persons over 65 years of age. The problems addressed in these three studies are the following.

Meta-Analysis

Do studies of antiplatelet therapy show that antiplatelet treatment of patients with transient ischemic attacks prevents stroke?

Decision Analysis

Should HIV-infected intravenous drug users be screened for tuberculosis and treated with isoniazid based on the results of the PPD test or should they all be given isoniazid without first screening with a PPD test?

Cost-Effectiveness Analysis

What is the cost of pneumococcal vaccination in persons over 65 per year of life saved?

3.2 DEVELOPING A STUDY PROTOCOL

3.2.1 Overview

A protocol is the blueprint for conduct of the study. It also serves as a permanent record of the original study objectives and of the study methods and procedures. A study protocol should be prepared before the study begins, and it should be freely available for others to review after the study has been completed.

   For all three types of study, the protocol should have a section on objectives, background, information retrieval, data collection, and analysis. The specific elements of each of these sections, listed in Table 3-1, differ between the three study types.

3.2.2 Objectives

The protocol should begin with a statement of the main objectives of the study. The statement should be concise and specific. For decision analysis, the statement of objectives should describe the utility measure for the analysis. For cost effectiveness analysis, the statement of the main objectives should describe the perspective of the analysis.

EXAMPLES:

Meta-Analysis

Problem: Do studies of antiplatelet therapy show that treatment of patients with transient ischemic attacks prevents stroke?

Objective: To identify all randomized trials of antiplatelet therapy in patients with transient ischemic attacks and do a meta-analysis to determine whether these studies show that antiplatelet treatment prevents stroke.

Table 3-1 Elements of a study protocol for meta-analysis, decision analysis, and cost-effectiveness analysis

Section	Meta-Analysis	Decision Analysis	Cost-Effectiveness Analysis
Objectives	State main objectives Specify secondary objectives	State main objectives Specify secondary objectives Specify utility measure and explain choice	State main objectives Specify secondary objectives State perspective of the analysis
Background	Brief review	Brief review	Brief review
Information retrieval	Describe overall strategy Specify MEDLINE search terms Explain approach to unpublished reports and "fugitive" literature	Give sources and explain these choices	Give sources and explain these choices
Data collection	Describe procedures for abstracting data from publications (blinding, reliability checks, handling of missing data) Describe quality rating scheme and procedures for assessing quality	Specify sources or procedures for collecting data on utility measure Describe how expert opinion will be solicited	Specify source of cost data
Analysis	State methods for estimating variance for individual studies State model to be used and explain choice Specify approach to missing data Describe how quality rating scheme will be used in the analysis	Specify sensitivity analyses	Specify sensitivity analyses
Other		Describe any human subjects considerations	Describe any human subjects considerations State discount rate and explain choice
Appendices	Copies of data collection forms	Copies of data collection forms Copies of informed consent documents	Copies of data collection forms Copies of informed consent documents

Decision Analysis

Problem: Should HIV-infected intravenous drug users be screened for tuberculosis and treated with isoniazid based on the results of the PPD test or
should they all be given isoniazid?

Objective: To do a decision analysis that uses information on the likelihood
of being PPD-positive, the likelihood of anergy in HIV-infected persons,
the likelihood of developing tuberculosis, and the risks and effectiveness
of isoniazid in order to compare PPD screening followed by isoniazid
treatment of persons with a positive PPD with isoniazid treatment without PPD testing on expected life expectancy in HIV-infected intravenous
drug users.

Cost-Effectiveness Analysis

Problem: What is the cost of pneumococcal vaccination per year of life
saved?

Objective: To do a cost-effectiveness analysis of pneumococcal vaccination
from the perspective of Medicare.

If there are secondary objectives, these should also be presented in the protocol.
Subgroup analysis is frequently a secondary objective of meta-analysis studies.
Describing plans for secondary analyses in the protocol is important because the
subgroup analyses are documented as being a priori aims of the study.

3.2.3 Background

The protocol includes a few paragraphs on the background of the study with
enough citations to the literature so that the reader can form an accurate picture
of the state of knowledge on the topic of the study. Key references to prior work
should be cited; a complete review of the world's literature on the topic of the
study is not necessary.

3.2.4 Information Retrieval

Systematic procedures for searching the literature are the centerpiece of meta-
analysis, and the protocol for meta-analysis should describe these procedures in
detail. The protocol section on literature search should begin with a statement of
whether the meta-analysis will include only published studies or whether an
attempt will be made to also identify and include unpublished studies. All of the
computer databases that will be searched should be listed. If there are computer
databases with possibly relevant information that will not be searched, a rationale
for this decision should be provided. The exact search terms and the search algorithm for each computer database should be presented in the protocol. If the literature search will include an attempt to retrieve unpublished studies, this section
of the protocol should include a description of the procedures for identifying these
studies.

"Fugitive" literature is a term applied to studies published in documents that either are difficult to identify because they are not abstracted or are difficult to retrieve because of their limited circulation. The fugitive literature includes dissertations, conference proceedings, and some government reports. If studies published in the fugitive literature will not be included, a rationale for this decision should be presented in the protocol.

Decision analysis and cost-effectiveness analysis also rely heavily on published data as a source of information on probabilities. When known ahead of time, the sources that will be used to estimate probabilities should be stated. If these are unknown, the criteria for deciding how to identify relevant literature should be specified. Where there are alternative sources of information about probabilities, the criteria for choosing among the alternatives should be stated explicitly.

Just as in a decision analysis, a protocol for a cost-effectiveness analysis should include information on how probabilities will be assessed. It should also give the source for data on cost. Alternative sources of cost data and the reasons they were not chosen should be described. The protocol should provide a rationale for use of charge data to estimate cost when this is the plan of the study. The discount rate used in the analysis should be given along with the rationale for the choice of the discount rate. The protocol should state whether or not benefits will also be discounted.

3.2.5 Data Collection

For meta-analysis, the section on data collection should begin with a statement of the inclusion and exclusion criteria that will be used to decide which of the studies identified in the literature search will be abstracted. The procedures for abstracting data should be presented in detail. If the abstractor is to be blinded to, for example, the journal of publication or to the results when abstracting data on the methods, the method for ensuring blinding should be described. If no attempt to blind the abstractor will be made, this fact should be stated explicitly. If reliability checks will be made, the frequency for these should be specified along with a plan for using the information. The rules for choosing from among several estimates of effect should be given. The procedures for handling missing data should be described.

If studies will be rated on quality, the protocol should describe how the rating scheme was developed. The procedures for abstracting information on quality should be specified. If the information is going to collected with the abstractor blinded to any aspect of the study, the manner of ensuring blinding should be described.

For decision analysis, the protocol should state what utility measure will be used. The source of information on the utility measure should be given. If utility will be assessed by primary data collection as part of the decision analysis, the procedures for doing this should be described.

When "guesses" of probabilities will be used in place of estimates from the literature or empiric studies, the protocol should define who will make the guess, and it should give the framework for the guess. If experts will be the source of opinion on probability measures, the criteria for selection of the experts should be given.

If data on cost will be gathered by primary data collection, the procedures for collecting these data should be described.

3.2.6   Analysis

For a meta-analysis, the analysis section of the protocol should state whether the pooling of data will be based on a fixed-effects model or a random-effects model. The methods that will be used to estimate the variance of studies that do not provide variance in a form usable in the chosen model should be given with appropriate references.

For decision analysis and cost-effectiveness analysis, plans for sensitivity analysis should be given, to the extent that these can be anticipated.

3.3   ACQUIRING RESOURCES

Planning a study that uses meta-analysis, decision analysis, or cost-effectiveness analysis should include a realistic assessment of resource needs for the study. All three types of studies can require a substantial investment in personnel resources. It is easy to underestimate these needs.

> *EXAMPLE:* A computer literature search for meta-analysis of menopausal estrogen replacement therapy and breast cancer (Dupont and Page 1991) identified 565 published studies that were potentially relevant to the topic of the meta-analysis. If it were necessary to retrieve each article from a library to determine whether it was eligible for inclusion in the meta-analysis and if retrieval took only 10 minutes per article, 94 hours of personnel time would be required to retrieve all 565 articles. If as many as 80% of articles identified through the computer search could be eliminated from consideration based on reading the title and the abstract, which might take only 2 minutes per article, the task of identification and retrieving relevant articles would still take almost 34 hours of personnel time.

After articles are retrieved, information from them must be abstracted. Abstraction also is time-consuming, especially when it is done systematically onto structured forms and when an attempt is made to blind the abstractor to various aspects of the study. Computer resources will be needed for most analyses that use these three methods. It also may be desirable to purchase special-purpose software for some of the studies, and this may be costly.

3.4   PROCURING ADMINISTRATIVE APPROVALS

Virtually all clinical trials, surveys, and case-control and cohort studies require some administrative approvals. These approvals include such things as approval to recruit patients from a clinic, approval to have access to medical records, and

Table 3-2   Examples of types of administrative
approvals that might be needed in a synthetic study

Meta-analysis
    Approval to use unpublished data
Decision analysis
    Approval to review medical records
    Approval to survey patients about treatment preferences or
    values
    Approval to use administrative databases
Cost-effectiveness analysis
    Approval for access to Medicare or other cost databases
    Approval to review patient billing records

approval to use a certain desk for the project recruiter. Fewer administrative approvals are necessary for a synthetic study, but failure to anticipate those that are necessary may delay the study or even prevent its completion. Some of the kinds of administrative approvals that might be needed in a meta-analysis, a decision analysis, or a cost-effectiveness analysis are listed in Table 3-2.

Most studies that use meta-analysis, decision analysis, or cost-effectiveness analysis will not fall within the scope of review by an institutional review board for the protection of human subjects. However, whenever a study will gather information that would potentially allow individuals to be identified, it should be submitted for review and approval to the committee early, to avoid delays in study completion. For example, a study that will use medical records to estimate probabilities for a decision analysis when information on the relevant probabilities is not available from a published source requires review by an institutional review board. It is not uncommon to review billing records for a sample of patients with a condition of interest to gather data to estimate cost in a cost-effectiveness analysis, and this also requires review and approval. Access to some large administrative databases may require institutional review board approval.

EXAMPLE: The most important effect of cigarette smoking during pregnancy on the infant is its effect on birth weight. Smoking increases the chances of being low birth weight and preterm (before 37 weeks) by a factor of about 1.3, and it increases the likelihood of being low birth weight and term by about 4.0. The cost of caring for an infant who is low birth weight and preterm is much higher than the cost of caring for one who is low birth weight but term, because preterm infants are much sicker. In our study of the cost effectiveness of smoking cessation programs during pregnancy (Shipp et al. 1992), published information on the cost of hospitalization for low birth weight infants who are preterm and term was not available, and administrative databases that would allow this classification were not available. To obtain it, a special study involving primary data collection was carried out. Delivery logs at two hospitals were reviewed to identify infants who were low birth weight. The infants were classified according to whether they were preterm or term by reviewing the mothers' medical records. Copies of the billing records of these infants were then retrieved and abstracted.

Since linkage of information from the delivery logs with the maternal medical records and with the billing records required individuals to be uniquely identified, it was necessary to submit this study to to review by an institutional review board for the protection of human subjects.

4

Information Retrieval

Because of the rapidity of growth of knowledge, the task of identifying and retrieving information for a study that uses any of these methods is formidable. The problem of information retrieval has received a great deal of attention in meta-analysis. Development of the field of meta-analysis has highlighted the importance of a systematic approach to information retrieval as a means of obtaining information that is free from bias. The importance of a systematic and unbiased approach to information retrieval is not restricted to meta-analysis, since most of the probabilities used in a decision analysis are estimated based on existing published information.

Section 4.1 describes an overall strategy for comprehensive retrieval of published information for a specific topic. Section 4.2 describes a strategy for computerized searches of MEDLINE as a method for retrieving published studies, while Section 4.3 describes other computer-stored databases of information. Section 4.4 discusses the limitations of searches of computerized databases and some ways to overcome these limitations. Section 4.5 describes the problem of publication bias and some ways to assess and handle it.

4.1 Overall Strategy for Retrieving Information on Published Studies

Decisions about what studies should be eligible for a meta-analysis and what information should be used to estimate probabilities or costs must be scientifically defensible and free of bias. Only when all information on the given topic is identified is the investigator in a position to achieve this goal. Ideally, the search for

Table 4-1   Usual steps for comprehensive retrieval of published information on a
specific topic

Step 1 Search personal files

Step 2 Do computerized literature search of computer-stored databases
    A. Search titles and abstracts and eliminate obviously irrelevant ones
    B. Retrieve remaining articles
    C. Review articles systematically and eliminate those that don't contain needed
       information

Step 3 Review reference lists of articles to find new articles

Step 4 Retrieve newly identified articles and review them for relevance

Step 5 Consult experts

information for a synthetic study would identify all of the relevant information.
In practice, the retrieval of information for all three types of studies is usually
limited to retrieval of published information.

Table 4-1 lists the steps generally used to retrieve published information for a
meta-analysis. These steps also should be followed to retrieve information for
decision analysis or cost-effectiveness analysis whenever there is a need to estimate
a probability and this estimate is going to be based on published data.

The first step in comprehensive retrieval of information is almost always a
search of the personal files of the investigator or knowledgeable colleagues to identify materials that are already in hand. This search is followed by a computerized
search of one or more computer databases, virtually always including MEDLINE.
The titles of publications identified in the computer search and their abstracts,
when available, are scanned to eliminate articles that are obviously irrelevant. The
full text of the remaining articles is then retrieved. These articles are read quickly,
and those that clearly are not relevant are put aside. The remaining publications
are then systematically reviewed to determine whether they are eligible for the
meta-analysis based on predetermined criteria for eligibility.

The reference lists of the articles that contain useful information are reviewed
to identify publications on the same topic that have not yet been identified. When
new articles are identified by this procedure, these articles are retrieved and the
process of reading them to ascertain whether they are eligible continues.

Simultaneously, other information sources are explored (Table 4-2). The
abstracts of published doctoral dissertations should be searched by computer if
there is any possibility that information on the topic may be contained in them.
A hand search of the journals that are known to publish papers on the subject
matter of the analysis may identify studies that otherwise remain unknown. If
information prior to 1966 would be of interest, a hand search of *Index Medicus*
should be done. This can be very important for some topics.

EXAMPLE: Mahomed and Hytten (1989) did a meta-analysis of randomized trials of the effect of routine iron supplementation on anemia in
pregnancy. Table 4-3 lists the seven studies that provided information on
the effect of iron on anemia defined as a hemoglobin less than 10–10.5 mil-

Table 4-2   Other strategies for retrieval of published studies

Hand search *Index Medicus* for period prior to MEDLINE (1966)

Hand search journals known to publish material in subject area of interest

Table 4-3   Studies included in meta-analysis of effect of routine iron supplementation on anemia at 36–40 weeks, year of publication, journal, and total number of subjects

Reference	Year of Publication	Journal	Number of Subjects
Holly	1955	*Obstetrics and Gynecology*	149
Pritchard and Hunt	1958	*Surgery, Gynecology, and Obstetrics*	123
Morgan	1961	*Lancet*	65
Chisholm	1966	*Journal of Obstetrics and Gynaecology of the British Commonwealth*	144
Fleming et al.	1974	*Medical Journal of Medical Science*	60
Batu et al.	1976	*Israel Journal of Medical Science*	55
Taylor et al.	1982	*British Journal of Obstetrics and Gynaecology*	46

Source: Mahomed and Hytten (1989); table references cited there.

ligrams per deciliter at 36–40 weeks of gestation. Three of the seven publications, involving 334 of the 639 subjects who had ever participated in such studies, were published before 1966 and would not have been identified in a computer search of MEDLINE.

When a list of all of the articles identified by the foregoing methods is complete, it is often submitted to an expert for review. The expert is asked to identify other publications that may contain information on the topic of interest. These articles are retrieved and reviewed systematically to determine eligibility for the meta-analysis.

4.2   COMPUTERIZED SEARCHES OF MEDLINE

4.2.1   Overview

MEDLINE is a bibliographic database that is the computerized counterpart of *Index Medicus.* It is the primary source of information on publications in the biomedical literature. MEDLINE contains information on publications in over 3500 journals indexed in *Index Medicus* and additional material from about 600 journals not indexed in *Index Medicus.* Along with its backfiles, MEDLINE covers *Index Medicus* for the period from 1966 to the present.

MEDLINE does not contain everything that is published in the journals that are indexed. Editorials, commentaries, and letters to the editor are subject to special rules about indexing. Indexing of letters in the early years of the database is very incomplete. Published abstracts are selectively indexed.

MEDLINE is not a full-text database. That is, the complete text of publications is not available in computer-stored form. Rather, for each indexed publication, MEDLINE contains the title, the authors, and the source of publication; the author abstract, if one is available; and a number of "medical subject heading" (MeSH) terms. The MeSH terms are chosen from a limited vocabulary, and they are assigned to each published article by a professional indexer working under a set of highly structured rules.

4.2.2   Access to MEDLINE

Access to MEDLINE is available through commercial telecommunications companies, such as Telenet and Tymenet, or through a software and telecommunications package, Grateful Med, that was developed by the National Library of Medicine. Both linkages with the MEDLINE database require a computer with a modem. The user must pay to search MEDLINE and MEDLARS using both commercial telecommunications software and Grateful Med, and it necessary to establish an account with the National Library of Medicine before the databases can be accessed and searched.

Many libraries provide access to the most recent years of MEDLINE using computer terminals located in the library, and some universities make access to this version of MEDLINE available at remote sites such as faculty offices. There is often no charge for searching MEDLINE in this form, but not all years of the MEDLINE database are included in these library-accessible forms. MEDLINE is available in the form of a CD-ROM disk. As with many library-accessible forms, all years of MEDLINE are not included on the CD-ROM disk. Most information retrieval projects will eventually require access to all years of MEDLINE by computer linkage with the National Library of Medicine using a commercial telecommunications package or using Grateful Med.

4.2.3   Structure of MEDLINE

Indexing of publications in MEDLINE began in 1966. There are now millions of indexed publications in the MEDLINE database. Because of the large number of indexed publications, MEDLINE is organized into a number of files, each of which contains information on publications in two to three years. The file that contains the most recently published information is always called MEDLINE for the purposes of a search. Earlier files are called "backfiles." The backfiles are given names that include the earliest year of information in the file. For example, there is a backfile that includes information on publications in the years 1966 through 1971; it can be accessed with any of three names—M66, BACK66, B66.

To comprehensively identify publications for a period from the inception of MEDLINE to the present, the most current file, called MEDLINE, and each of the backfiles must be searched separately.

4.2.4 Search Strategies

Publications in MEDLINE can be identified by searching the words that appear in the title or the abstract (textword search), by exact title, by author, by source, or by MeSH term.

4.2.4.1 Textword Search

Conceptually, the easiest kind of search to do is a textword search. In a textword search, all publications in the MEDLINE database that use the specified word either in the title or in the abstract are identified. Unless the searcher uses a special procedure that identifies "roots," a textword search is a search for the exact word. That is, a search for the textword "estrogen" will not identify publications that use "estrogens" in the title or abstract.

There are a number of problems with textword searches. First, textword searches often identify a large amount of material. It can be impossible to review all of the identified material for possible relevance. At the very least, reviewing the material is costly. Second, in spite of the fact that textword searches tend to identify large amounts of material, they do not necessarily identify all of what is relevant.

> *EXAMPLE:* To identify publications for a meta-analysis of studies of stroke and estrogen replacement therapy, it would be possible to start with a textword search of MEDLINE that would identify all studies of with "estrogen" in the title and abstract. To examine this search strategy, a textword search of the 1986–1988 MEDLINE was done. This search identified 4696 publications with the word "estrogen" in the title or the abstract. Sorting through all of these publications to try to determine whether they contained information on the subject of stroke would be a formidable undertaking. The cost of printing the title, the author, the source, and the abstract for all of the publications identified in this textword search so that this information could be perused at leisure to assess relevance would be very large. In spite of all the work and expense that might go into review of the material from this textword search, it would not identify a publication by Boysen et al. (1988) that contained information on the relative risk of stroke in women using estrogen replacement therapy.

4.2.4.2 Title Search

MEDLINE can be searched to identify a publication with a specific title. In a title search, the exact title of the publication that is sought must be known. For this reason, title searches are not a useful method for identifying material whose content is not known with certainty.

4.2.4.3 Source Search

MEDLINE can also be searched by source. That is, all of the articles published in a given journal and indexed in MEDLINE can be identified. To do a source search, it is necessary to know the MEDLINE abbreviation for the journal that is to be searched. With some exceptions, the abbreviations of journals are assigned according to a set of rules defined in publications of the National Library of Medicine.

> *EXAMPLE:* A journal with the word "American" in the title includes the abbreviation "Am" as part of the MEDLINE abbreviation; a journal with the word "Journal" in the title includes "J" in the MEDLINE abbreviation; and a journal with the word "Epidemiology" in the title includes "Epidemiol" in the MEDLINE abbreviation. Putting these rules together yields "Am J Epidemiol" for the journal, *American Journal of Epidemiology.*

In a meta-analysis, a source search can help determine whether information is likely to be published in a journal with a particular target audience. For example, the journal *Gastroenterology* frequently publishes studies about gallbladder disease. After identifying all of the publications in *Gastroenterology* using MEDLINE, the titles and abstracts can be reviewed quickly to identify information relevant to a meta-analysis, a decision analysis, or a cost-effectiveness analysis that was not otherwise identified.

4.2.4.4 Author Search

MEDLINE can be searched by author. That is, all of the publications with a certain person as one of the authors can be identified. In an author search, it is necessary only to specify the last name of the author in question. For example, all of the publications written by a person with the last name "Smith" can be identified. By also specifying the first and last initial of the author, the number of publications by authors with the specified last name identified will be limited. For example, all publications by "AB Smith" can be identified; this is obviously fewer than the number of publications with any "Smith" as author.

An author search may be useful in comprehensive retrieval of information for a given topic when it known that certain people regularly publish studies in a given subject area.

> *EXAMPLE:* To identify publications that used meta-analysis for this book, a search of the MEDLINE database for publications by the author TC Chalmers in the years 1980–1990 was done. TC Chalmers was known to publish regularly about meta-analysis. This search identified 15 articles that were pertinent to this book.

Author searches present some problems. First, an author search on a common name such as Smith may identify a lot of irrelevant material, and sifting through it may be time-consuming. If the author search is done by specifying both initials in an attempt to cut down on the amount of irrelevant material, publications by that author that did not use both initials will be missed. Second, if a name is misspelled in the original publication, it will not be identified in the author search.

4.2.4.5 Subject Search

Publications in MEDLINE are reviewed by expert indexers and assigned "medical subject headings," called MeSH headings. The number of MeSH terms assigned to a publication varies. Most articles are assigned between 8 and 15 MeSH terms. MeSH terms are chosen from a limited vocabulary of terms that was developed by the National Library of Medicine and is periodically updated. MeSH terms are assigned to publications by professionals who are highly trained in the use of the system.

The assignment of MeSH terms to publications is a special feature of MEDLINE that allows the searcher to identify relevant published material even when the author might not have used the subject term in the title or the abstract. This feature allows searches of this large and complex database to be focused and specific. It may greatly reduce the amount of material that needs to be retrieved while simultaneously achieving completeness in the information retrieval process.

EXAMPLE: Returning to the problem of identifying studies of estrogen replacement therapy and stroke described, a search of the MED86 backfile using the MeSH terms "cerebrovascular disorder" and "risk factors" was done. This search identified 200 potentially relevant publications, compared with 4696 publications based on the textword search. In contrast with the textword search, the MeSH search identified the Boysen et al. (1988) publication that had relevant information.

Searching for publications according to their assigned MeSH terms requires knowledge of the system for indexing articles and access to a MeSH dictionary. The MeSH vocabulary undergoes constant revision and the rules for assigning MeSH terms are periodically updated, making MeSH searches of the entire database from 1966 to the present a complex undertaking. Effectively searching MEDLINE using MeSH requires training and experience. The National Library of Medicine offers courses designed to train professionals to search MEDLINE, and many university libraries offer courses with the same goal. Taking a formal course in use of the system or consultation with a professional librarian trained in the use of MEDLINE is almost always necessary to use the MeSH system effectively. Further detail on the use of the MeSH vocabulary to search MEDLINE is beyond the scope of this book.

4.3 OTHER COMPUTER-STORED DATABASES

MEDLINE is just one of several computer-stored databases operated by the National Library of Medicine and subsumed under the term MEDLARS (Medical Library Information Retrieval System). Other computer databases that might contain information that would be of use in a meta-analysis, a decision analysis, or a cost-effectiveness analysis are described briefly in Table 4-4. Procedures for searching these databases are not covered in this book.

Table 4-4　Computer-stored bibliographic databases other than MEDLINE that might be searched in a meta-analysis, decision analysis, or cost-effectiveness analysis

Database	Description of Contents
AIDSLINE	AIDS-related records from journal articles, government reports, meeting abstracts, special publications, and theses (1980 to present)
CANCERLIT	Cancer literature from journal articles, government and technical reports, meeting abstracts, published letters, and theses (1963 to present)
Dissertation Abstracts Online	American and Canadian doctoral dissertations (1861 to present)
TOXLINE	Effects of drugs and other chemicals from journal articles, monographs, theses, letters, and meeting abstracts

4.4　LIMITATIONS OF COMPUTERIZED SEARCHES OF COMPUTER-STORED DATABASES

4.4.1　Overview

The availability of MEDLINE and other computer-stored bibliographic databases has greatly aided the identification of potentially relevant published material for studies involving meta-analysis, decision analysis, and cost-effectiveness analysis. Access to computer bibliographic databases has by no means solved the problem of information retrieval. The databases are incomplete. Indexing and search algorithms are imperfect. Practical constraints on retrieval of articles can lead to incompleteness of even perfect searches. Most important, computer-stored bibliographic databases contain only part of the literature of medicine.

4.4.2　Incomplete Databases

No single computerized database covers all periodicals, even for a defined, broad subject area like medicine. MEDLINE, for example, contains information on original research reported in less than one-third of all biomedical journals.

Journals that are not included in MEDLINE are described as highly specialized journals on topics considered to be of limited interest, journals of low circulation, and journals in which articles have not been peer-reviewed. These criteria do not ensure that important data have not been published in them.

> *EXAMPLE:* In the meta-analysis of published studies of iron supplementation and anemia that was described in Table 4-3, the study by Batu et al., which was published in the *Israel Journal of Medical Science,* would not have been identified in a computer search of MEDLINE because publications in this journal were not indexed in the year of publication.

4.4.3 Imperfect Search Algorithms

Even if the computerized data sources contained every journal, it is not always possible to search the database in such a way that every pertinent article is retrieved. Developing a search algorithm that identifies all of the pertinent material that appears in a computer-stored database takes considerable expertise even for simple problems. Even algorithms developed by experts fail.

> *EXAMPLE:* The National Perinatal Epidemiology Unit at Oxford University has compiled a register of controlled trials in perinatal medicine using a variety of methods, including contact of individual investigators, hand searches of the perinatal journals, and perusal of meeting reports. This publication list was considered the "gold standard" in a comparison with a search of the MEDLINE database by Dickersin et al. (1985). The investigators chose two topics—neonatal hyperbilirubinemia and intraventricular hemorrhage. Without knowledge of the contents of the register, a medical librarian experienced in computer searches developed a search strategy designed to identify all pertinent published articles reporting the results of randomized trials of neonatal hyperbilirubinemia or intraventricular hemorrhage indexed in the MEDLINE database for the period of operation of the MEDLINE database, 1966–1983.
>
> In the register, there were 88 English-language publications on neonatal hyperbilirubinemia that were confirmed to be randomized trials and 29 on intraventricular hemorrhage. The MEDLINE search identified only 28 English-language publications on neonatal hyperbilirubinemia and only 19 on intraventricular hemorrhage, as shown in Table 4-5.
>
> Table 4-6 shows that 81 of the 88 published trials of neonatal hyperbilirubinemia registered in the Perinatal Database and 27 of the 29 published trials of intraventricular hemorrhage registered were indexed in MEDLINE and potentially retrievable. Thus, the failure to retrieve some of the articles was strictly due to the search strategy, which was imperfect despite the fact that it was done by an expert.
>
> The MEDLINE search identified some pertinent articles that had not been included in the "gold standard" register.

The ability of a search algorithm to identify all of the pertinent literature can be improved by consultation with an expert searcher.

> *EXAMPLE:* In the comparison of the Perinatal Trials Database with the MEDLINE search described above (Dickersin et al. 1985), an "amateur" searcher, Dr. Thomas Chalmers, developed a search strategy independently. His strategy identified only 17 studies of neonatal hyperbilirubinemia and only 11 of intraventricular hemorrhage, as shown in Table 4-5.

When a set of publications known to be relevant is available, it may be useful to review the indexing terms for the set of articles to identify terms that would have allowed the articles to be retrieved in MEDLINE. Search algorithms should

Table 4-5   For two topics, the number of randomized trials eligible for a meta-analysis that were registered in perinatal trials database and the number found in MEDLINE search

Topic	Number of Studies Registered in Perinatal Database	Number of Studies Found in MEDLINE Search	
		Expert Searcher	Amateur Searcher
Neonatal hyperbilirubinemia	88	28	17
Intraventricular hemorrhage	29	19	11

Source: Dickersin et al. (1985).

Table 4-6   For two topics, the number of published studies in the perinatal trials database and the number that were indexed in MEDLINE

Topic	Number of Published Studies in the Perinatal Database	Number of Those Published Studies Indexed in MEDLINE
Neonatal hyperbilirubinemia	88	81
Intraventricular hemorrhage	29	27

always be checked carefully to be sure that the terms used in them apply in all years of the database.

4.4.4   Imperfect Indexing

The success of searches that are based on use of index terms depends greatly on the accuracy of indexing and on the ability of the medical subject heading terms, or other indexing procedure, to capture relevant information. Indexing itself is limited by the low quality and inaccuracy of the descriptions of the research in some source documents.

4.4.5   Practical Constraints on Retrieval

Publications in foreign-language journals are included in most computerized databases such as MEDLINE. Their titles are usually provided in English. However, deciding whether an article is relevant by perusing the translated title is difficult. In MEDLINE, an English-language abstract is available only if the author provided one and it was published in the journal. Thus, if a publication cannot be judged to be definitely relevant or definitely not relevant based on its title, there may not be other clues to relevance. Retrieval of the full text of foreign-language journals can be difficult, and getting a translation of an article is expensive, if it can be obtained at all. For all of these reasons, the tendency is to limit information retrieval for research synthesis to studies published in English. This practical constraint may seriously limit the completeness of information retrieval based on computerized databases.

> *EXAMPLE:* For a meta-analysis designed to estimate the rate of neural tube defects in children born to women who had taken clomiphene citrate, we obtained from Merrell Dow, the pharmaceutical company that manufactures and distributes the drug worldwide, a complete set of all published studies reporting data on follow-up of women who had conceived after taking clomiphene citrate. Complete translations of all works published in languages other than English that had been done by professional translators were provided by the company. After reviewing each of the 187 articles provided by the company, 56 were considered eligible for the meta-analysis by virtue of presenting information on pregnancy outcome, including presence or absence of birth defects, for at least 5 women who had conceived after taking clomiphene citrate. Of the 56 studies that met inclusion criteria for the study, results of 21 (37.5%) were not available in English and would have been excluded from the meta-analysis if we had not access to the Merrell Dow publications database and their professional translations.

Whenever possible, pertinent data published in languages other than English should be identified.

4.4.6 Fugitive Literature

Government reports, book chapters, the proceedings of conferences, and published dissertations are often called "fugitive" literature because the material published in them is difficult to identify and because the documents or their contents may be difficult to retrieve. Studies published in conference proceedings, as book chapters, and in government reports are not identified in searches of MEDLINE and most other computer databases. Experts may be an especially good source of information about publications in the fugitive literature. Writing to persons who have published in the field of interest and specifically asking them if they have published relevant material in a report or a book chapter or if they know of any reports or book chapters with information may be fruitful.

It is possible to find appropriate published dissertations by searching *Dissertation Abstracts,* and this is a step in any literature search that strives to be entirely complete.

The amount of material on a given topic that may appear in the fugitive literature is highly variable, depending on the topic.

> *EXAMPLE:* Elbourne, Oakley, and Chalmers (1989) did a meta-analysis of the effect of social support on the rate of low birth weight. Table 4-7 lists the nine randomized trials of the topic that they identified after an exhaustive search for studies, along with the place of publication, if published. Of the five published studies on this topic, only three were published in periodicals that would have been identified in a search of MEDLINE.

In contrast, Table 4-8 shows the number of studies that were published

Table 4-7 Studies included in meta-analysis of effect of social support on likelihood of low birthweight, the place of publication, the number of subjects, and the estimated relative risk

Reference	Place of Publication	Number of Subjects	Estimated Relative Risk
Blondel et al.	Unpublished	152	1.43
Oakley et al.	Unpublished	486	0.84
Heins et al.	Unpublished	1346	0.91
Spencer and Morris (1986)	*Prevention of Preterm Birth* (book)	1183	1.05
Elbourne et al. (1987)	*British Journal of Obstetrics and Gynaecology*	273	0.82
Lovell et al. (1986)	*Pediatric and Perinatal Epidemiology*	197	0.48
Reid et al. (1983)	*Report to Health Services Research Committee*	155	0.60
Olds et al. (1986)	*American Journal of Public Health*	308	2.09
Dance	Unpublished	50	0.72

Source: Elbourne, Oakley, and Chalmers (1989); table references cited there.

Table 4-8 Number of eligible studies published as abstracts or doctoral dissertations or in books, government reports, or conference proceedings for six meta-analyses chosen for the presumed rigor of identification of publications

Reference	Topic	Total Number of Eligible Studies	Number Published as Abstracts or Doctoral Dissertations or in Books, Reports, or Proceedings
Collins, Yusuf, Peto (1985)	Diuretics during pregnancy[a]	11	0
Law, Frost, Wald (1991)	Salt reduction and blood pressure[a]	70	1
Littenberg (1988)	Aminophylline for acute asthma[a]	13	1
Longnecker et al. (1988)	Alcohol and breast cancer[b]	16	0
Stampfer and Colditz (1991)	Estrogen and coronary heart disease[b]	32	2
Yusuf et al. (1985A)	Fibrinolytic therapy in acute myocardial infarction[a]	33	3

[a]Meta-analysis of randomized clinical trials.

[b]Meta-analysis of observational studies.

in the fugitive literature for six meta-analyses that were selected because of the presumed rigor of identification of pertinent material. For two of the meta-analyses (Collins, Yusuf, Peto 1985; Longnecker et al. 1988), none of the studies included in the meta-analysis had been published in the fugitive literature. For each of the other five meta-analyses, the number of studies published in the fugitive literature was very small.

Unfortunately, it is not possible to know ahead of time whether the number of studies published in the fugitive literature will be large or small, and the attempt to identify them must be vigorous.

Unpublished studies are the ultimate example of fugitive literature. The existence of large numbers of unpublished studies has the potential to cause bias and is of serious concern. Publication bias is discussed in detail in the next section of this chapter.

4.5 PUBLICATION BIAS

4.5.1 Definition

The term "publication bias" is usually used to refer to the greater likelihood of research with statistically significant results to be submitted and published compared with nonsignificant and null results. More generally, publication bias is the systematic error induced in a statistical inference by conditioning on the achievement of publication status (Begg and Berlin 1988). Publication bias occurs because published studies are not representative of all studies that have ever been done.

4.5.2 Evidence for Publication Bias

Existence of a bias in favor of publication of statistically significant results is well documented (Sterling 1959; Simes 1986; Easterbrook et al. 1991). The most extreme example comes from the social science literature and is provided by Sterling (1959), who found that 97% of a series of consecutive articles that used significance testing published in the mid 1950's in four prestigious psychology journals reported results that were statistically significant. His findings are shown in Table 4.9.

In the medical literature, Easterbrook et al. (1991) documented less extreme, but nonetheless serious, bias in favor of publication of statistically significant results. Examining the publication status of 285 analyzed studies for which institutional review board approval had been obtained between 1984 and 1987 at Oxford, the investigators found 154 studies had statistically significant results and 131 did not. Of the 154 studies with statistically significant results, 60.4% had been published, whereas only 34.4% of the studies that did not have statistically significant results had been published. These findings are shown in Table 4.10.

Simes (1986) showed that a conclusion based on meta-analysis about the effect of an alkylating agent alone compared with combination chemotherapy on sur-

Table 4-9   Number and percentage of articles using significance testing that
reported statistically significant ($p < 0.05$) results by journal

Journal	Year	Total Reports Using Tests of Significance	Reports That Were Statistically Significant[a] N	%
Experimental Psychology	1955	106	105	99.1
Comparative and Physiological Psychology	1956	94	91	96.8
Clinical Psychology	1955	62	59	95.2
Social Psychology	1955	32	31	96.9
All		294	286	97.3

[a]Reject H_0 with $p \leq 0.05$.
Source: Sterling (1959).

Table 4-10   Publication status in 1991 for 285 analyzed studies reviewed by the
Central Oxford Research Ethics Committee in 1984–1987

Publication Status	Study Result Statistically Significant[a] N	%	Not Statistically Significant[b] N	%
Published	93	60.4	45	34.3
Presented only	38	24.7	31	23.7
Neither published nor presented	23	14.9	55	42.0
Total	154	100.0	131	100.0

[a]$p < 0.05$.
[b]$p \geq 0.05$.
Source: Easterbrook et al. (1991).

vival in patients with advanced ovarian cancer would be different depending on
whether the meta-analysis was based on published studies or on studies registered
with the International Cancer Research Data Bank. Table 4-11 shows that when
only published studies were used, there was overall a statistically significant
increase in median survival in patients treated with combination chemotherapy.
When a group of studies noted in a registry at their initiation, before the results
were known, was used in the meta-analysis, there was no significant advantage of
treatment with combination chemotherapy.

There are no comparable concrete examples of bias toward publication of null
results, as opposed to statistically significant results, for any topic. Begg and Berlin
(1988), however, speculate that historically a bias toward publication of null
results may have characterized the study of asbestos and cancer. When there are
adverse financial or regulatory consequences of a positive result, a bias in favor of
publication of null or negative results is a theoretical possibility.

Table 4-11 Results of meta-analysis of published and registered studies of treatment with an alkylating agent alone compared with combination chemotherapy in patients with advanced ovarian cancer[a]

Results	Published Studies ($N = 16$)	Registered Studies ($N = 13$)
Median survival ratio[b]	1.16	1.06
95% confidence interval	1.06–1.27	0.97–1.15
p value	0.02	0.24

[a]Adjusted for sample size.

[b]Median months of survival in patients treated with combination chemotherapy/median survival in patients treated with alkylating agent alone.

Source: Simes (1986).

It is common to attribute publication bias to editorial policies that favor publication of positive results and to bias of journal reviewers against negative results. Dickersin, Min, and Meinert (1992) did a follow-up of studies that were either approved or ongoing in 1980 by the two institutional review boards that serve the Johns Hopkins Health Institutions. As in the study of Easterbrook et al. (1991), completed studies with statistically significant results were more likely to have been published than studies with nonsignificant results. Over 90% of the unpublished studies had not been submitted for publication, and only 6 of the 124 unpublished studies had been submitted for publication and rejected by a journal.

4.5.3 Effects of Publication Bias on Decision Analysis and Cost-Effectiveness Analysis

Publication bias is a serious concern in decision analysis and cost-effectiveness analysis as well as in meta-analysis, since these methods also rely on published information to derive the probabilities of various events. Often in decision analysis and cost-effectiveness analysis, the assumption about the overall effectiveness of the intervention being studied derives from an aggregation of published studies. Thus, publication bias has the potential to affect the conclusions based on a decision analysis.

EXAMPLES: Table 4-12 presents the results of a decision analysis that compared treatment of patients with advanced ovarian cancer with an alkylating agent alone or with combination chemotherapy (Simes 1985). Therapy with combination chemotherapy has more serious side effects than therapy with an alkylating agent alone. When the difference in survival associated with combination chemotherapy was assumed to be 1.08, a value that is close to the value found in the meta-analysis based on all registered studies described above (Simes 1986), 9 of 9 women would favor the alkyl-

Table 4-12   Results of a decision analysis: Number of women who would favor alkalating agent alone over combination chemotherapy according to assumptions about median survival ratio by method used to assess utility of treatment

Utility Method	Median Survival Ratio[a]			
	1.08[b]	1.23[c]	1.58	1.75
Time trade-off	9/9	3/9	1/9	0/9
Standard gamble	9/9	2/9	1/9	0/9

[a]Median months survival in patients treated with combination chemotherapy/median months survival in patients treated with alkalating agent alone.
[b]Estimate close to result of meta-analysis of all registered studies.
[c]Estimate close to result of meta-analysis of published studies only.
Source: Simes (1985).

ating agent over combination chemotherapy. When the difference in survival with combination chemotherapy was assumed to be 1.23, a value close to the value found in the meta-analysis of all published studies (Simes 1986), only 2 or 3 of 9 women would favor the alkylating agent.

The effect of publication bias on decision analysis and cost-effectiveness analysis has not received much attention in the literature describing the conceptual basis for the methods. Decision analysis and cost-effectiveness analysis rely more and more on meta-analysis to estimate probabilities for both decision analysis and cost effectiveness, putting the results of the analyses on firmer scientific footing.

4.5.4   Solutions to the Problem of Publication Bias

The problem of publication bias will be solved completely only when investigators submit and editors accept all well-conducted studies of important questions irrespective of the statistical significance of their results. Until that time, there are three choices: ignore the problem; attempt to retrieve all study results, whether published or unpublished; or use statistical or quasi-statistical methods to assess or overcome it.

4.5.4.1   Ignore the Problem

Because the existence of publication bias is so well documented, it is impossible to ignore. At a minimum, the potential for publication bias to explain a study finding should be acknowledged explicitly. Documented examples of publication bias all show a tendency for preferential publication of statistically significant results. In the face of a null result, it may be useful to point this out. When unpublished studies have not been sought, it may be useful to discuss the incentives and disincentives for publishing results that are not statistically significant.

Table 4-13   Some existing registries of clinical trials

Registry	Subject
Oxford Database of Perinatal Trials	Randomized trials in perinatal medicine
International Committee on Thrombosis and Haemostasis	Multicenter randomized trials in thrombosis and hemostasis
International Cancer Research Database	Protocols in cancer chemotherapy

4.5.4.2   Attempt to Retrieve All Studies

Retrieval of information from all studies, not just published studies, is an appealing and theoretically ideal solution to the problem of publication bias. When there is a registry of trials, then it is possible to write to all investigators in the registered trials and obtain the results of the trial, even when it has not been published. This task may be laborious, and, when investigators fail to respond to queries about the trial, the possibility of bias remains.

Retrieval of information from all studies of a given topic may be infeasible, because the data on the topic may not even have been analyzed. For example, a meta-analysis of coffee drinking and coronary heart disease based on unpublished as well as published studies would require identification of all the studies that have collected data on coffee and coronary heart disease, requiring review of original questionnaires for a large number of studies.

Identification of unpublished as well as published studies is most realistic when studies are registered in their planning stages or when they begin. Several registries of clinical trials, described in Table 4-13, already exist. The pressure to solve the problem of publication bias for clinical trials is likely to lead to an increase in the number and scope of these registers in the future. When a register of trials is available, it should be used.

Registers of nonexperimental studies have not yet been created. The obstacles are formidable, since many nonexperimental studies are based on secondary analysis of data collected for a purpose other than the original goal of the study.

Editorial policies that place higher value on studies that report "statistically significant" associations should be abandoned. Greater reliance on interval estimation and decreased emphasis on significance testing should continue to be encouraged.

4.5.4.3   Statistical and Quasi-Statistical Approaches

Several statistical and quasi-statistical approaches to assessing and dealing with publication bias have been described. The statistical approaches developed to date lack a firm footing in formal statistical theory and make assumptions that are dubious or untenable. These statistical approaches are not recommended, for reasons that are discussed in detail in Chapter 7.

Light and Pillemer (1984) describe a graphical technique for assessing the possibility of publication bias that is quasi-statistical. The effect measure is plotted on the horizontal axis and the sample size on the vertical axis. In the absence of pub-

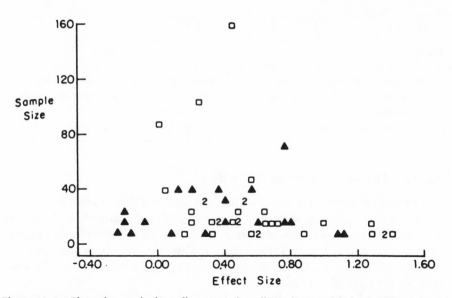

Figure 4-1 Plot of sample by effect size for all studies, published (squares) and unpublished (triangles), of psychoeducational interventions and hospital stay in surgical patients. The plot has the expected funnel shape. (Reproduced with permission from Light RJ and Pillemer DB: *Summing Up: The Science of Reviewing Research.* Harvard University Press, Cambridge, Massachusetts, 1984, p. 69.)

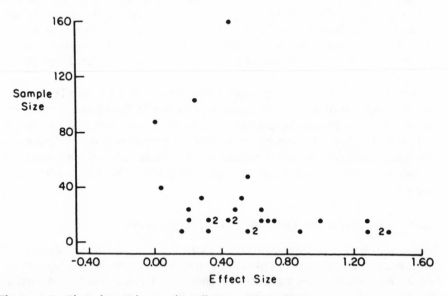

Figure 4-2 Plot of sample size by effect size for published studies only. The left corner of the funnel, which should contain the results of small studies with negative and null results, is missing. (Reproduced with permission from Light RJ and Pillemer DB: *Summing Up: The Science of Reviewing Research.* Harvard University Press, Cambridge, Massachusetts, 1984, p. 68.)

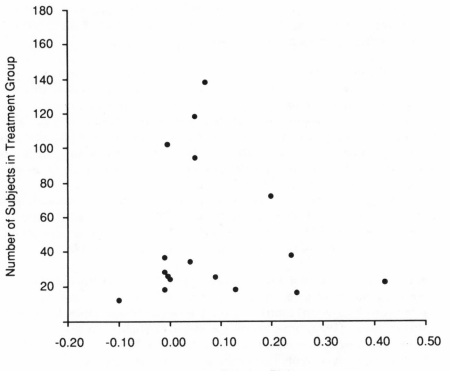

Figure 4-3   Plot of sample size by logarithm of the relative risk for published studies only. The plot is not selectively missing small studies with negative and null results even though there is other evidence to suggest that there is publication bias for this topic. Data are from Simes (1987).

lication bias, because of sampling variability, the graph should have the shape of a funnel that is viewed sideways with the large opening down and the tip pointed up and centered on the true effect size. If there is bias against the publication of null results or results showing an adverse effect of the treatment, the left corner of the pyramidal part of the funnel will be distorted or missing.

EXAMPLE: Light and Pillemer (1984) provide an example of distortion of the expected funnel shape of a plot of sample size and a measure of effect in a situation where there was evidence of publication bias. The data were drawn from a meta-analysis of the effect of psychoeducational interventions on hospital length of stay in surgical patients (Devine and Cook 1983). The meta-analysis of published studies showed a statistically significant association of the intervention with a reduction in hospital length of stay, whereas an analysis of published studies plus studies reported in dissertations showed no statistically significant association and an estimated effect of 0.0. A plot of sample size against effect size for the whole group of published and unpublished studies, shown in Figure 4-1, was in the expected funnel shape.

A plot of sample size against effect size for the published studies only, shown in Figure 4-2, shows that the left corner of the funnel, which should contain small studies reporting negative or null results, is missing.

In medical applications, a funnel plot should be scaled to ensure that negative effects and positive effects are equally spaced. For studies that measure effects on a ratio scale (relative risk or odds ratio), this can be accomplished either by plotting the measure on a logarithmic scale or by taking the logarithm of the measure and plotting it on a linear scale. For these studies, the measure of statistical variability of the measure of effect should be plotted in preference to the sample size, since total sample size for case-control and cohort studies does not provide information about the variability of the effect measure.

The sensitivity of funnel plots as a method for detecting the existence of publication bias has not been assessed systematically. When a funnel plot is distorted, publication bias should be suspected. But even when a funnel plot does not provide clear-cut evidence of publication bias, this possibility cannot be ruled out.

EXAMPLE: Figure 4-3 shows a funnel plot for the published studies of an alkylating agent alone compared with combination chemotherapy in patients with advanced ovarian cancer that was discussed earlier. For these studies, Simes (1987) provided clear evidence of publication bias, and the funnel plot would be expected to demonstrate this. For these data, it is difficult to say with certainty that the funnel plot is distorted.

5

Data Collection

Meta-analysis, decision analysis, and cost-effectiveness analysis all rely on abstraction of information from written, usually published, reports. Decision analysis and cost-effectiveness analysis may involve, in addition, primary data collection. Data collection in synthetic studies should adhere to the same standards of quality assurance that guide data collection in other types of research studies. It is no more acceptable to collect data on scraps of paper in a meta-analysis, a decision analysis, or a cost-effectiveness analysis than it would be to collect data on scraps of paper in a clinical trial.

Section 5.1 describes the overall goals of the process of data collection in synthetic studies. Section 5.2 discusses reliability of data collection and ways of enhancing reliability of collection of data from written reports. Section 5.3 discusses the concept of validity. Section 5.4 discusses bias in the collection of data from written reports and ways to minimize it.

5.1 OVERALL GOALS

The main goal of data collection in any study is the collection of information that is reliable, valid, and free of bias. In addition, the data collection process should create a permanent record of the information that was used in the study so that others can review it. The requirement for a permanent record of data used in the study applies equally to information collected for use in meta-analysis, decision analysis, and cost-effectiveness analysis. In the absence of a permanent, well-organized record of the results of data collection, it is impossible for others to be sure even that the work was done. When information has been collected from written reports, the record of data collection makes it possible to recheck the information used in the study, and it facilitates bookkeeping.

5.2 RELIABILITY

5.2.1 Definition

A measure is reliable if it is free of measurement error. Reliability is assessed by determining the extent to which the measurement is consistent—whether the measurement is the same when made a second time. Reliability is a concept that applies both to individual items of data and to scales composed of an aggregation of items.

There are three pertinent measures of reliability. Intrarater reliability is a single rater's consistency in rating an item when it is presented more than once. Test–retest reliability is consistency of values when rated twice by the same person, with some interval of time between the independent ratings. Interrater reliability is consistency in rating by different persons.

Reliability is a matter of concern in abstraction of information from written reports. This includes the abstraction of information about the number of subjects and the year of publication, as well as information on the summary measure of effect and information used to rate the quality of the study. In decision analysis and cost-effectiveness analysis, reliability of information about the probabilities used in the analysis should be a concern, as well as the reliability of measures of utility and cost.

5.2.2 Reliability of Data Abstracted from Written Reports

Superficially, it would seem that information abstracted from written publications ought to be inherently reliable. After all, the information is written down in black and white. Reliability of information abstracted from written reports cannot be assumed for several reasons. First, data collection from written reports is subject to simple human error. The numbers given in the publication may be recorded incorrectly. These simple human errors lead to unreliability. Second, in most written reports, measures of the same quantity are cited more than once. For example, a measure of the overall association of the treatment with the outcome may be presented in the abstract, in the narrative section of the results, in a table, and again in the discussion. A simple instruction to record the measure of association may lead one abstractor to record the number reported in the abstract, whereas another abstractor might record the number reported in the results section.

EXAMPLE: In a publication reporting on mortality follow-up of women in the Walnut Creek Contraceptive Drug Study (Petitti, Perlman, Sidney 1987), the estimated relative risk of death from cardiovascular disease in "ever users" of noncontraceptive estrogens compared with "never users" after adjustment for age and other cardiovascular disease risk factors was cited as 0.5 in the abstract but was given as 0.6 in the relevant table in the text of the article. The discrepancy between the two citations arose because a correction to the abstract made to the galley proofs was not incorporated into the published manuscript. This fact cannot be known to a person abstracting information on cardiovascular disease from the written report, and it leads to unreliability in the measure of risk of cardiovascular disease in non–contraceptive users abstracted from this publication unless an instruction is given to consistently abstract information from one or the other section of the published paper.

Last, the information desired from a written report may not be presented exactly in the form in which it is needed. In this case, the abstractor must make assumptions about the material presented in the text. The assumptions made from one reading of the text to the next by the same abstractor, or by different abstractors reading the same text, may not be identical, and the data may be unreliable for this reason.

There are few formal evaluations of the reliability of data abstraction for meta-analysis. The most thorough examination of the issue of reliability for meta-analysis is a study of interrater reliability by Stock et al. (1982). A random sample of 30 documents was selected from a list of sources of information on the topic of the meta-analysis. A statistician and two postdoctoral educational researchers coded 27 items from the 30 studies. The items included judgments such as whether or not a study should be coded, as well as calculations from the original data sources such as the median age of subjects in the study. Correlations between pairs of coders ranged from 0.57 for quality of the study to 1.00 for median age and the standard deviation of age. The low level of reliability for some items suggests that specific procedures to maximize reliability are necessary.

5.2.3   Enhancing Reliability

5.2.3.1   General Procedures

Table 5-1 adapted from Stock et al. (1982), lists several ways to enhance the reliability of abstraction of information from written reports. Most of these procedures apply to data collection in all kinds of studies, not just studies that rely on abstraction of data from written reports. The general procedures for abstracting information include development of data collection forms before the study begins, pilot testing of the forms, and revision of the forms based on the results of the pilot test.

When information on a topic is available only from a single source, as is often the case in a decision analysis, a single form may need to be developed for a single piece of information. Alternatively, a single form that documents the source of data on each of the probabilities may be developed.

When data will be abstracted from multiple reports, a detailed book of abstracting instructions must be prepared. This book should include rules for each

Table 5-1 Ways to enhance the reliability of data collection for studies involving abstraction from published literature

Develop and pilot test forms before the study begins

Develop a detailed book of abstracting instructions and rules for each form.

Provide training to abstractors based on the rule book and forms

Assess interabstractor reliability

Revise the rule book and forms and retrain abstractors as needed

Develop procedures for adding new abstractors

Encourage abstractors' involvement in discussions and decisions about abstracting instructions and rules

Source: Stock et al. (1982).

of the forms and for all items on the forms that are not entirely self-explanatory. The abstracting book should be updated when decisions to change rules, procedures, or forms are made.

Abstractors should be formally trained. In training sessions, as described by Stock et al. (1982), the principal investigator should meet with the abstractor, describe each item, and go through the process of abstracting a report with the abstractor observing. Graduate students and fellows who are doing their own data abstraction also need to be trained.

The early forms abstracted by the newly trained abstractor should be reabstracted to assess formally the reliability of the abstraction process. A sample of forms should be periodically abstracted as part of ongoing quality control. When new abstractors are hired, they should also be trained and the reliability of their work monitored closely during the early stages of the study.

Stock and colleagues (1982) emphasize the importance of involvement of abstractors in discussions about the instructions and rules for abstraction. Implicit is the notion that those closest to the data are most knowledgeable about the problems and their solutions.

For some items of information, it may be useful to have two or more independent abstractions. Discrepancies in the judgments of the two readers can then be adjudicated by a master reader or by a consensus of the study investigators meeting as a committee.

EXAMPLE: Littenberg (1988) did a meta-analysis of aminophylline treatment in severe, acute asthma. To assign a quality score to each study, he developed a nine-point scale in which the presence or absence of nine characteristics of a well-designed study was assessed for each study. To determine the quality score, each report was reviewed by two board-certified internists with training in clinical epidemiology.

5.2.3.2 *Format and Organization of Data Collection Forms*

Attention to the format and organizational structure of the standardized data collection form is important. Good forms used to abstract information from written

Table 5-2 Important features of forms for abstraction from published literature

---

Each item should have a unique identifying number

Each item should require a response, regardless of whether information is or is not available.

Skip and stop instructions should be clear; they should be printed on the form

The language should be simple

The layout should be pleasant; it should facilitate data entry by standardizing the location of check items

The pages should be numbered sequentially and the total number of pages in the form should be indicated on each page

The sequence of items should be logically related to the format of most scientific publications

---

Source: Meinert (1986).

reports have features in common with all good survey instruments and data collection forms. Table 5-2, modified from Meinert (1986), lists the features that are especially important in the design of forms for abstraction from written reports.

Each item on the form should have a unique identifying number. This facilitates data entry and data analysis. Each item should require a response, regardless of whether information on the item is or is not available. This allows missing information to be distinguished from an incomplete abstraction. Skip and stop instructions should be clear, and these instructions should be printed on the form. Doing this makes it possible for others to understand how data collection flows. The language used on the form should be simple. Use of esoteric terms, unnecessary words, and double negatives should be avoided, for reasons that are obvious. The pages of the forms should be numbered sequentially, and the total number of pages in the form should be indicated on each page. In this way, missing pages can be spotted and the abstractor will know when the form has been completed. The sequence of items on the form should be logically related to the material to be abstracted. For example, items related to information that is likely to be presented in the methods section should precede items related to information likely to appear in the results section.

5.3 VALIDITY

Information collected by data abstraction from a written report or in a survey of persons is valid if it measures what it is intended to measure. In a meta-analysis, the information on the estimate of effect is valid to the extent that the original study results are valid. Other aspects of data collection in synthetic studies pose more tractable questions with regard to validity.

In measuring study quality, a valid measure is a measure that adequately represents the quality of the study. In measuring utilities for a decision analysis, a valid measure of utility is one that measures the desirability or acceptability of the outcome. Validity and assessment of validity in development of quality rating scales are discussed in Chapter 6. Chapter 11 discusses the validity of measures of utility.

5.4 BIAS

5.4.1 Definition and Origins

Bias is systematic error that distorts the truth. While measurement error makes it more difficult to find true associations, bias leads to erroneous conclusions. It is very difficult to detect biased data collection from reading the results of a study. When data collection is biased, the problem is almost impossible to overcome analytically.

Bias in data collection from written reports can arise in many ways, some of them quite subtle. The abstractor who believes strongly that one treatment is better than another may select the data from the report most favorable to this position. The knowledge that the study is "positive" or is "negative" may lead this biased abstractor to search harder for information that would ensure that the study is excluded from or included in a meta-analysis. Knowledge that a study was published in a prestigious journal may cause the abstractor to rate the paper more highly on measures of quality, when these measures are being used, and this will ultimately result in bias.

The possibility of biased data abstraction seems to be especially great when the abstractor is a professional who has stated in public an opinion on the topic or who has a financial stake in the results. When the investigator is in this position, it is probably wise to recruit or hire an independent person to do the data abstraction.

5.4.2 Reducing Bias in Data Collection from Written Reports

5.4.2.1 Blinding

Blinding of the abstractor to aspects of the publication that might influence the data abstraction is the best way to eliminate bias in data abstraction. The aspects of the written report that are most likely to influence the abstractor are the authors and their institution, the journal of publication, and the funding source (Chalmers et al. 1981). To blind the abstractor, each study can be assigned a code number. A heavy black marking pen can be used to cross out information that would allow the author, the journal, and the funding source to be identified (Chalmers et al. 1981).

> *EXAMPLE:* In the quality rating that Littenberg (1988) did for his meta-analysis of aminophylline treatment in severe, acute asthma, identifying information—authors, titles, journal, institution, and country of origin— were removed from the report before they were given to the internists for abstraction.

Many journals are identifiable by their type style, layout, or page size. The distinctive three-column format of the *Journal of the American Medical Association* and its affiliated journals and the exceptionally small type size of the *British Medical Journal* are examples. The distinctiveness of these features may make it impossible to completely blind experienced authors and investigators to the journal of publication. Experienced investigators and subject matter experts are often so familiar with publications that they recognize them when they are being abstracted. Most authors will recognize their own work. The inability to blind experienced investigators and subject matter experts is another argument for recruiting or hiring an independent person to do data abstraction, when funds are available.

5.4.2.2   Task Separation

Knowledge of the results of the study may influence the perception of the methods. For this reason, abstracting information from the methods section separate from the results section may help reduce bias. Two separate forms, one for methods and one for results, can be abstracted by the same person on separate occasions. Alternatively, two abstractors might be employed, and the task of abstracting the methods section assigned to one of them and that of abstracting the results section to the other (Chalmers et al. 1981).

6

Advanced Issues in Meta-Analysis

The principles of protocol development, comprehensive identification of information, and data collection that were described in earlier chapters apply to each of the three methods of quantitative synthesis. This chapter begins the presentation of advanced issues pertinent only to meta-analysis.

Having developed a strategy for identifying studies using the methods described in Chapter 4 and data collection forms as described in Chapter 5, the next steps in a meta-analysis are to define eligibility criteria for the meta-analysis, to apply these criteria to the selection of studies, and to select the data from the studies that are deemed eligible for the meta-analysis. It may be desirable to rate study quality. These topics are covered in this chapter. Chapter 7 describes statistical methods for meta-analysis of data from comparative studies and Chapter 8 describes other statistical aspects of meta-analysis.

Section 6.1 discusses the goals, timing, and process of defining eligibility criteria and determining the eligibility of studies for the meta-analysis and gives an overview of eligibility criteria. Sections 6.2 through 6.8 discuss the rationale for the eight basic eligibility criteria. Section 6.9 discusses the choice of estimates of effect from studies with more than one estimate. Section 6.10 describes how to develop a quality rating system for meta-analysis.

6.1 DEFINING ELIGIBILITY CRITERIA AND DETERMINING ELIGIBILITY OF INDIVIDUAL STUDIES

6.1.1 Overall Goals

The overall goals of the process of defining eligibility criteria for a meta-analysis are to ensure reproducibility of the meta-analysis and to minimize bias. The selection of studies is reproducible if another person arrives at the same conclusion about which studies will be included in the meta-analysis. If the eligibility criteria are well described, the selection of studies will be reproducible. Bias in the selection of studies for the meta-analysis can arise when selection is influenced by knowledge of the results or by other aspects of the study design. Defining eligibility criteria reduces bias by assuring that decisions about eligibility are systematic.

Defining eligibility criteria of studies for a meta-analysis is analogous to defining eligibility criteria for entry to a randomized trial or for inclusion of subjects in a case-control or cohort study. The decision to include or exclude a person from a study of any design should not be arbitrary, and it should not be based on convenience. It should be based on sound scientific reasoning. Similarly, the decision to include or exclude a study from a meta-analysis should be based on sound scientific reasoning.

6.1.2 When to Define Eligibility Criteria and Determine Eligibility

Eligibility criteria for the meta-analysis should be defined *before* the abstraction of data from the studies begins. The criteria should be described in a protocol that has been prepared in advance of the conduct of the study. Developing eligibility criteria ahead of time and describing them in a protocol documents the intent of the analyst and protects the meta-analysis from allegations that the choice of studies in the final analysis was influenced by investigator bias. Documentation is particularly important when the investigator has a previously stated position on the topic, when the investigator has published one or more studies on the topic that are in conflict with other studies, and when the results of the meta-analysis have beneficial or adverse monetary consequences for the investigator or for the sponsor of the research.

It is especially important to determine whether a study is eligible for the meta-analysis before carrying out the statistical analysis of the studies. Doing this minimizes bias. When the effect estimates from each study are known, it may be possible to pick and choose from among them and arrive at just about any conclusion.

EXAMPLE: Table 6-1 presents estimates of the relative risk of stroke in users of estrogen replacement therapy that Grady et al. (1992) identified as eligible for a meta-analysis of the effects of hormone replacement therapy on disease. The overall estimate of relative risk of stroke in estrogen users based on all of the studies considered eligible by Grady et al. is 0.96 (95%

Table 6-1   Estimates of the relative risk of stroke in users of estrogen replacement therapy for studies considered eligible for meta-analysis

Reference	Stroke Endpoint	Estimated Relative Risk (95% Confidence Interval)
Pfeffer (1976)	Fatal and nonfatal first stroke	1.12 (0.79–1.57)
Petitti et al. (1979)	Hospitalized fatal and nonfatal stroke	1.19 (0.67–2.13)
Rosenberg et al. (1980)	Occlusive stroke	1.16 (0.75–1.77)
Adam, Williams, Vessey (1981)	Fatal subarachnoid hemorrhage	0.64 (0.06–6.52)
Wilson, Garrison, Castelli (1985)	Fatal and nonfatal stroke or TIA	2.27 (1.22–4.23)
Bush et al. (1987)	Fatal stroke	0.40 (0.01–3.07)
Boysen et al. (1988)	Fatal and nonfatal first stroke	0.97 (0.50–1.90)
Henderson, Paganini-Hill, Ross (1991)	Fatal occlusive stroke	0.63 (0.40–0.97)
Stampfer et al. (1991)	Fatal and nonfatal stroke	0.97 (0.65–1.45)
Finucane et al. (1992)	Fatal and nonfatal first stroke	0.65 (0.45–0.95)
Summary relative risk: all studies		0.96 (0.82–1.13)
Summary relative risk: studies published before 1986		1.25 (0.99–1.57)
Summary relative risk: studies published in 1986 or later		0.76 (0.61–0.94)

Source: Grady et al. (1992); table references cited there.

C.I. 0.82–1.13). After inspecting the estimates of relative risk, it is easy to see that omission of the studies published before 1986 would eliminate all of the studies that found an increased risk of stroke in estrogen users. The summary estimate of the relative risk of stroke in users of postmenopausal estrogen after excluding the five studies published before 1986 is 0.76 (95% C.I. 0.61–0.94).

Studies done before 1986 might be of lower quality than later studies, and a cogent argument for excluding them might be made on these grounds. If the argument is an a priori argument, then excluding them is justifiable. If the argument is made after inspecting the relative risk estimates, the critical reader might suspect investigator bias.

Equally plausible is an argument that concern about the risk of postmenopausal estrogen stemming from the adverse publicity about hormones and vascular disease, which was common in the late 1970s and early 1980s, led physicians to prescribe estrogen to an increasingly healthy group of women. Based on this reasoning, one might exclude the studies published in later years. The summary estimate of the relative risk of stroke based on studies published before 1986 is 1.25 (95% C.I. 0.99–1.57). Again, if the decision to exclude later studies is an a priori decision based on sound reasoning about the likelihood of selection bias for estrogen use, it is reasonable to base the meta-analysis only on the early studies. If the decision follows inspection of the data, it is difficult to refute the allegation of investigator bias.

6.1.3 The Process of Assessing Eligibility

The process of determining whether studies are eligible for inclusion in the meta-analysis should be systematic and rigorous. Each potentially eligible study should be scanned in a specified order. The determination of whether or not a study meets the predetermined eligibility criteria should be made by personnel who have been trained and who work from a set of written instructions. The goal of each of these procedures is to ensure reliability of determinations of eligibility and to minimize bias.

The reasons why a study is deemed ineligible should be recorded, and a log of ineligible studies needs to be maintained (Chalmers et al. 1981). Studies that are directly pertinent but are not included in the meta-analysis should be cited in the published report on the meta-analysis, and the reasons for rejecting them should be presented. These procedures allow others to assess the completeness of the literature review process and the accuracy of the application of the stated eligibility criteria. Keeping a log of ineligible studies is analogous to keeping a log of patients who were screened for a clinical trial but were not enrolled, and the goals are the same.

It is desirable to make decisions about eligibility with the abstractor blinded to the results of the study and probably also to the source of publication and the authors (Chalmers et al. 1981). Blinding has the same purpose as blinding in experiments—it minimizes bias. Without blinding, it may be tempting, for example, to include studies published in prestigious journals even if they do not meet the eligibility criteria.

When the number of studies being considered is small enough, the decisions about eligibility should be reviewed independently by another abstractor. It may be especially important to re-review rejected studies. When there are discrepancies between the judgments of two independent reviewers, they should be resolved using a predetermined procedure, such as a conference or adjudication by an expert.

6.1.4 General Eligibility Criteria

Table 6-2 lists seven basic considerations about eligibility that should be addressed in almost all meta-analyses. First, the designs of eligible studies should be speci-

Table 6-2 Basic considerations in defining eligibility for a meta-analysis

Study designs to be included

Years of publication or study conduct

Languages

Choice among multiple publications

Restrictions due to sample size or follow-up

Similarity of treatments and/or exposure

Completeness of information

fied. Second, the inclusive dates of publication, presentation, or conduct of studies eligible for the meta-analysis should be given. Third, the eligibility of studies whose results are not available in English should be addressed. Fourth, the criteria for choosing among results of multiple publications from the same study population should be defined. Fifth, any restrictions on eligibility due to sample size or length of follow-up should be stated. Sixth, eligibility or ineligibility based on the similarity of treatments or exposure should be considered. Last, the eligibility or ineligibility of incomplete and unpublished reports should be addressed.

6.2 STUDY DESIGN

The average effect of a new treatment has generally been found to be larger in nonrandomized than in randomized studies (Chalmers, Block, Lee 1972; Sacks, Chalmers, Smith 1983; Wortman and Yeatman 1983; Colditz, Miller, Mosteller 1989; Miller, Colditz, Mosteller 1989). When both randomized and nonrandomized studies are available for a topic, estimates of effect size should be made separately for the randomized and the nonrandomized studies.

When all available studies are nonexperimental, it is difficult to make a rule about similarity of study design as an eligibility criterion for the meta-analysis. The study design that is likely to yield the most valid conclusion may be different for different topics. There may be conflict between the results of studies of different designs, and it may not be possible to say which study design yields the correct conclusion.

> *EXAMPLE:*　Table 6-3 shows estimates of the relative risk of osteoporotic fracture in female smokers according to whether the study was a case-control study or a cohort study. All of the case-control studies found a higher risk of osteoporotic fracture in smokers, whereas all of the cohort studies found a relative risk of 1 or less.
>
> In almost all of the case-control studies, response rates were higher in cases than in controls. Response rates in smokers are lower than in nonsmokers in most population-based surveys, and response bias may explain the higher relative risk observed in case-control studies of smoking and fracture. Case-control studies are also subject to recall bias, which would spuriously increase the estimates of relative risk of fracture in smokers. Considering these issues, it would seem that cohort studies yield the more correct conclusion. On the other hand, information about smoking in many cohort studies is defined at entry to the study and is not updated thereafter. The absence of an association of smoking with osteoporotic fracture in cohort studies may be due to nondifferential misclassification of smoking status and a bias to the null that would result from this failure.

In the interest of completeness, it may be prudent to consider studies of all designs to be eligible for a meta-analysis of nonexperimental studies. If there is statistical evidence of heterogeneity in the estimate of effect, the possibility of study design as an explanation for the discrepancy in estimates of effect can be examined.

Table 6-3 Estimated relative risk of osteoporotic fracture in female smokers by study type

Reference	Estimated Relative Risk
Case-control studies	
Daniell (1976)	4.2
Aloia et al. (1985)	3.2
Paganini-Hill et al. (1981)	1.96[a]
Kreiger et al. (1982);	1.29
Kreiger and Hildreth (1986)	
Williams et al. (1982)	6.5[b]
Alderman et al. (1986)	13.5[c]
Cohort studies	
Jensen (1986)	0.7
Hemenway et al. (1988)	1.0[d]
Holbrook, Barrett-Connor, Wingard (1988)	1.1

[a]>11 cigarettes/day.

[b]Hip fracture in average weight smoker compared with obese nonsmoker.

[c]Hip fracture in thin smoker compared with obese nonsmoker.

[d]≥25 cigarettes/day.

Source: United States Department of Health and Human Services (1990); table references cited there.

EXAMPLE: Stampfer and Colditz (1991) did a meta-analysis of studies of estrogen replacement therapy and coronary heart disease. There was strong statistical evidence of heterogeneity ($p < 0.001$) in the estimate of relative risk derived from the entire group of 31 studies that were eligible for the meta-analysis. When the relative risk of coronary heart disease was estimated separately by study design, differences were found. Hospital-based case-control studies and prospective cohort studies without internal controls, which were judged to have the greatest likelihood of bias, yielded the highest and the lowest estimates of relative risk (Figure 6-1). Estimates derived from prospective studies with internal controls and cross-sectional studies of angiographically determined coronary artery disease, which were considered to be least prone to bias, were consistent with each other and, when combined, yielded a summary estimate of relative risk of coronary heart disease of 0.50 (95% C.I. 0.43–0.56).

6.3 INCLUSIVE DATES OF PUBLICATION

Because of the availability of MEDLINE as a method for identifying published literature, it has become common to define 1966, the date when MEDLINE began, as the starting date for identification of studies eligible for a meta-analysis. The fact that MEDLINE began in 1966 is not an adequate justification for defining 1966 as the starting date for eligibility for a meta-analysis. The inclusive dates of publication should be chosen based not simply on convenience but on consid-

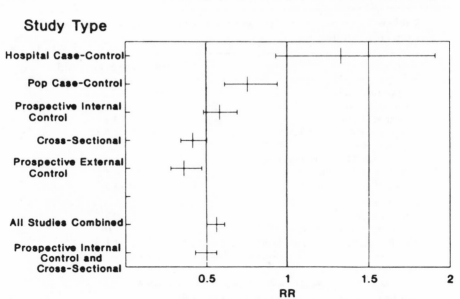

Figure 6-1 Summary estimates of relative risk (RR) and their 95% confidence intervals for studies of postmenopausal estrogen use and coronary heart disease, by type of study design. (Reproduced with permission from Stampfer and Colditz, *Preventive Medicine,* 1991;20:58.)

eration of the likelihood of finding important and useful information during the period that is chosen.

> *EXAMPLE:* Table 4-3 showed the results of an exhaustive review done by Mahomed and Hytten (1989) to identify studies of the effect of routine iron supplementation during pregnancy. Eight studies were identified; only four of them were published in 1966 or later. Two of the three largest randomized trials of this topic were done before 1966. Of the 643 subjects who had participated in randomized studies, 337 had participated in studies whose results were published before 1966.

The meta-analysis should be as up-to-date as possible. The cutoff date for identification of eligible studies should be specified in the report on the meta-analysis so that material published after the cutoff date will not be assumed to have been missed in the literature search.

6.4 ENGLISH-LANGUAGE PUBLICATIONS

The potential problems with inclusion only of publications in English were discussed in Chapter 4. When only English-language publications will be eligible for the meta-analysis, the rationale for this decision should be stated.

6.5   MULTIPLE PUBLICATIONS FROM THE SAME STUDY POPULATION

Multiple published reports from the same study are very common. The summary estimates of effect from reports on the same study population are not independent, and including more than one estimate of effect from the same study population violates the statistical assumptions that underlie the procedures for aggregating data. Information from the same study population should contribute only once to the summary estimate of effect. Failure to exclude multiple reports from the same study population has the potential to cause bias in the summary estimate of effect.

> *EXAMPLE:* In the meta-analysis of stroke and estrogen replacement therapy described in Section 6.1.2, Grady et al. (1992) found information on the relative risk of stroke in users of estrogen replacement therapy in two publications based on follow-up of women living in Leisure World, in two publications based on women in the Walnut Creek Contraceptive Drug Study, and in two publications from the National Health and Nutrition Examination Survey (NHANES) Follow-up Study. Although the period of follow-up and the number of events differ, the estimates of effect are not independent, and information from each study should be included only once in computing the summary estimate of relative risk. Only one estimate of relative risk was used to derive the summary estimate of the relative risk of stroke in estrogen users described in Table 6-1. Table 6-4 shows estimates of the relative risk of stroke for all of the *publications* on stroke in estrogen users. In this table, information from multiple publications from the Leisure World cohort, the Walnut Creek Contraceptive Drug Study, and the NHANES cohorts is listed separately. The summary estimate of the relative risk of stroke for estrogen replacement therapy is 0.89 (95% C.I. 0.77–1.03) when based on all publications. The summary estimate of relative risk was 0.96 (95% C.I. 0.82–1.13) when only one report from each study population was used. Although the overall conclusion about the effect of estrogen on stroke risk is not different between the two analyses, use of all publications would overestimate the public health benefit of estrogen use for stroke.

6.6   RESTRICTIONS ON SAMPLE SIZE OR LENGTH OF FOLLOW-UP

Some of the statistical methods for meta-analysis are asymptotic methods. Asymptotic methods will tend to overestimate the precision of small studies (Greenland 1987). When the precision of a study is overestimated, it will carry too much weight in the meta-analysis. To avoid the problem of weighting small studies inappropriately in the meta-analysis, it is reasonable to make sample size an eligibility criterion for the meta-analysis. Small studies are excluded.

For some topics, the length of follow-up may influence the likelihood of

Table 6-4 Estimated relative risk of stroke in users of estrogen replacement therapy from all publications

Study Population	Reference[a]	Estimated Relative Risk (95% Confidence Interval)
Leisure World	Pfeffer (1976)	1.12 (0.79–1.57)
	Paganini-Hill, Ross, Henderson (1988)[b]	0.53 (0.31–0.91)
	Henderson, Paganini-Hill, Ross (1991)[b]	0.63 (0.40–0.97)
Northern California Kaiser-Permanente Health Plan	Rosenberg et al. (1980)	1.16 (0.75–1.77)
Oxford, England	Adam, Williams, Vessey (1981)[b]	0.64 (0.06–6.52)
Walnut Creek Study	Petitti et al. (1979)	1.19 (0.67–2.13)
	Petitti, Sidney, Perlman (1987)[b]	0.6 (0.2–2.2)
Framingham	Wilson, Garrison, Castelli (1985)	2.27 (1.22–4.23)
Lipid Research Clinics	Bush et al. (1987)[b]	0.40 (0.01–3.07)
Denmark	Boysen et al. (1988)	0.97 (0.50–1.90)
Nurses' Health Study	Stampfer et al. (1991)	0.97 (0.65–1.45)
National Health and Nutrition Examination Survey	Finucane et al. (1991)[b]	0.35 (0.14–0.88)
	Finucane et al. (1992)	0.65 (0.45–0.95)
Summary relative risk		0.89 (0.77–1.03)

[a]Table references are as cited in Grady et al. (1992).
[b]Studies of fatal stroke.

observing a true association. For example, a treatment for breast cancer probably will not have an effect on the likelihood of breast cancer recurrence in the first year after treatment. Including studies with only one year of follow-up of breast cancer patients in a meta-analysis of treatments for breast cancer would bias the estimate of the effect of treatment to the null. Similarly, there are many situations where exposure would not affect the risk of disease until after a latent period. For example, exposure to asbestos would not be expected to affect the risk of mesothelioma in the first 5 years after exposure. Including studies with less than 5 years of follow-up of persons exposed to asbestos in a meta-analysis of asbestos and mesothelioma would bias the result to the null. To avoid these problems, length of follow-up could be a criterion for eligibility for the meta-analysis.

Restrictions based on sample size or length of follow-up should be specified in advance and documented in the protocol.

An alternative to making study size or length of follow-up an eligibility criterion is to estimate effect with and without small studies or with and without studies with short follow-up or low-dose exposure. This is a kind of sensitivity analysis, which is discussed further in Chapter 13.

6.7 ELIGIBILITY BASED ON SIMILARITY OF TREATMENTS (OR EXPOSURES) OR OUTCOMES

One of the most critical decisions about eligibility for a meta-analysis is the decision about how similar the treatments (or exposures) and the outcome must be to pool them in the same analysis. These decisions are varied. They include deciding whether to consider together studies with different doses of the same drug; studies of chemically different drugs with the same mechanism of action; and studies of diseases in the same organ system but with a possibly different underlying pathophysiology.

> *EXAMPLES:* The Antiplatelet Trialists' Collaboration (1988) was a meta-analysis of randomized trials of antiplatelet drugs as treatment for cardiovascular disease. Four different antiplatelet drugs—aspirin, sulfinpyrazone, dipyridamole, ticlopidine—had been studied in various different randomized trials. Aspirin had been studied at more than eight different doses ranging from 5 to 1500 milligrams.
>
> Trials of long-term beta-blocker drugs after acute myocardial infarction and mortality reported by Yusuf et al. (1985B) included studies of propanolol, metoprolol, atenolol, sotalol, timolol, practolol, alprenolol, oxprenolol, and pindolol.
>
> The studies of stroke and estrogen replacement therapy used in the meta-analysis by Grady et al. (1992) described in Section 6.1.2 included studies of hospitalized occlusive stroke, fatal subarachnoid hemorrhage, hospitalized stroke (including both occlusive and hemorrhagic stroke) only, and fatal occlusive stroke only.

When the treatments evaluated in different studies do not have the same effect on outcome, including them in the meta-analysis may bias the meta-analytic summary estimate of effect. When a treatment has a strong effect on one outcome but not on another, including them may also bias the summary estimate of effect.

> *EXAMPLES:* In the meta-analysis of beta-blocker drugs after acute myocardial infarction by Yusuf et al. (1985B), the overall odds ratio of death in persons treated long-term with a beta-blocker drug was 0.77 (95% C.I. 0.70–0.85) when based on all trials. The odds ratio for death based on the 11 trials of long-term treatment with the beta blockers with intrinsic sympathomimetic activity (practolol, alprenolol, oxprenolol, pindolol) was 0.90 (95% C.I. 0.77–1.05). The odds ratio was 0.69 (95% C.I. 0.61–0.79) when based on the 14 trials of drugs without intrinsic sympathomimetic activity (metoprolol, atenolol, propranolol, sotalol, timolol). The inclusion of treatments with drugs with sympathomimetic activity, which appear not to have much effect on mortality, biased the overall summary estimate toward the null.
>
> The summary estimate of the relative risk of stroke in users of estrogen replacement therapy based on all of the studies in Table 6-1 was, as previously mentioned, 0.96 (95% C.I. 0.82–1.13). The summary estimate of rel-

ative risk based only on studies of fatal stroke is 0.50 (95% C.I. 0.39–0.82). There are different interpretations of this observation. It is possible that estrogen use protects only against fatal stroke. Alternatively, estrogen users may be different from nonusers in ways that influence the likelihood that they will die from stroke but not their chances of suffering a stroke. In either case, the conclusions differ depending on the outcomes that are used to define eligibility for the meta-analysis.

Including disparate treatments and outcomes in the same meta-analysis may result in overgeneralization of the results of the meta-analysis.

EXAMPLES: A conclusion about beta blockade based on a meta-analysis of all 24 trials of long-term treatment and mortality would be that beta blockade is beneficial. However, the analysis is more consistent with the conclusion that the benefit of long-term treatment is confined to use of beta blockers that are without intrinsic sympathomimetic activity.

A conclusion about the effect of estrogen use on stroke based on all studies of the topic is that it is not beneficial considering stroke risk. The analysis only of fatal stroke suggests that estrogen use may prevent death from stroke even if it does not prevent incident disease.

Decisions about eligibility based on the similarity of the outcome or the treatment must balance strictness, which enhances the homogeneity of the studies, against all-inclusiveness, which enhances completeness. Highly restrictive eligibility criteria tend to give the meta-analysis greater face validity. But criteria may be so restrictive and require so much homogeneity that they limit the eligible studies to only one or two studies, thus defeating one of the goals of meta-analysis as a method to increase statistical power. Nonrestrictive criteria may lead to the accusation that the meta-analysis "mixes apples and oranges." For example, given the differences in the underlying pathophysiology of various stroke syndromes, it seems unlikely that estrogen use would have the same effect on both hemorrhagic stroke and ischemic brain infarction. Including studies with both these stroke endpoints in the same analysis does not make sense based on an understanding of the pathophysiology of the diseases.

6.8 COMPLETENESS OF INFORMATION

6.8.1 Inclusion of Unpublished Reports, Abstracts, Brief and Preliminary Reports

Information from a study may be available only as an unpublished report from a meeting, as an abstract, which may or may not have been published, or in preliminary form, as a letter to the editor or a brief report. These types of reports are all examples of incomplete or summary reports. A decision must be made on whether studies with information available only in an incomplete or summary form are eligible for inclusion in the meta-analysis.

Incomplete reports usually describe sketchily the design, analysis, and results of the study. The sketchiness creates two problems. First, it may be impossible to determine from the material in the publication whether the study meets the eligibility criteria that have been applied to more complete reports. Second, the study may not have been subjected to peer review. These considerations argue for exclusion of incomplete reports from the meta-analysis.

Research in several fields shows that only about 35–40% of abstracts are followed by a full report within four to five years (Dudley 1978; Goldman and Loscalzo 1980; Meranze, Ellison, Greenhow 1982; McCormick and Holmes 1985; Chalmers et al. 1990). If published full reports are prepared only for positive or statistically significant studies, omitting incomplete reports has the potential to create bias. On the other hand, if studies that are not followed by a publication in full form are of poorer quality than studies that are followed by publication in full form, omission may improve the overall quality of the meta-analysis.

Chalmers et al. (1990) rated the quality of 16 controlled trials originally published in summary form, which included abstracts, brief reports, and letters to the editor. Only 36% were followed by publication of a complete report within four years. Table 6-5 shows that, in this series, publication as a full report was *not* associated with higher quality.

Because there are no firm data showing that publication in full form is related to quality and because publication in full form may selectively favor positive or statistically significant reports (see Chapter 4), it is probably wise to include abstracts and other incomplete reports in a meta-analysis if they meet other eligibility criteria. Alternatively, analysis with and without the results of the incomplete reports can be done. This strategy is discussed more in Chapter 13 on sensitivity analysis.

6.8.2 Inclusion of Studies with Incomplete Data

Reports of studies that meet all of the eligibility criteria for a meta-analysis may not present an estimate of effect size, and the raw data that would allow an estimate to be calculated may not be available in the study report. These studies cannot contribute to the summary estimate of effect. They are analogous to dropouts in a clinical trial and nonresponse in a case-control study. Based on this analogy, it is probably best not to consider these studies ineligible. Rather, the study should

Table 6-5 For controlled trials originally published in summary form,[a] percent published in full form by quality score assigned to study

Quality Score	Number of Trials	Trials Published in Full Form	
		N	%
High (6+)	10	3	30.0
Medium (3–5)	114	43	37.7
Low (<3)	52	18	34.6

[a]As an abstract, brief report, or letter to the editor.

Source: Chalmers et al. (1990).

be identified as eligible. It should appear in the table of eligible studies with a blank to indicate the absence of usable information. In this way, readers will recognize that these studies were not missed in the literature search. If the number of eligible studies that could not be used to estimate the summary measure of effect is large, the reader will be alerted to the possibility of bias.

6.9 CHOOSING ESTIMATES OF EFFECT WITHIN ELIGIBLE STUDIES

It is rare for studies to present one and only one estimate of effect. Crude and adjusted estimates, estimates of effect in several subgroups, and estimates of effect including and excluding subjects who did not comply with treatment or complete the trial are often reported. The presentation of numerous estimates of effect is a particular problem in reports of nonexperimental studies, because extensive multivariate analysis is often done.

> *EXAMPLE:* Table 6-6 presents the estimates of relative risk of fatal cardiovascular disease in estrogen users from the Lipid Research Clinics study (Bush et al. 1987), which is one of the studies included in the meta-analysis by Stampfer and Colditz (1991) that was described in Section 6.3. The results of eight different multivariate analyses were described in the publication. Although all of the estimates of the relative risk of disease in users of estrogen are less than 1.0 and all analyses yield the same overall conclusion about the association of estrogen use with cardiovascular disease, it is uncertain which estimate is the most appropriate one to include in the meta-analysis.

For a meta-analysis, one and only one estimate of effect from each eligible study should be used in the calculation of the summary estimate of effect, because using more than one would inappropriately weight studies with many estimates of effect. When all of the studies are randomized, controlled trials, the estimate used in the meta-analysis should be the estimate that is based on the "once randomized, always analyzed" rule. Alternative analyses that use estimates that take into account loss to follow-up or compliance can also be done, but these analyses should be interpreted cautiously, just as analyses of the data from the individual trials that are not intention-to-treat analyses should be interpreted cautiously.

For nonexperimental studies, rules to choose from the estimates should be established before the analysis begins. The rules should be documented in the protocol. Table 6-7 lists some rules that could be used to choose from among effect estimates in nonexperimental studies. The abstractor should be trained to apply the rules, and the reliability of application of the rules should be evaluated.

When some of the variables included in a multivariate model are not true confounders, the model will be overfitted. The precision of the estimate of effect is reduced in overfitted models, and the 95% confidence intervals for all variables in the model will be wider than for the appropriately fitted model. Many of the statistical methods for estimating effect in nonexperimental studies use estimates

Table 6-6 Estimates of the relative risk of fatal cardiovascular disease in estrogen users presented in one publication by Bush et al. (1987)

Estimated Relative Risk (95% Confidence Interval)	Subgroup	Variable Adjusted
0.34 (0.12–0.81)	All	Age only
0.37 (0.16–0.88)	All	Age, blood pressure, smoking
0.44 (0.19–1.03)	All	Age, blood pressure, smoking, total cholesterol
		Age, blood pressure, smoking, LDL cholesterol, HDL cholesterol
0.44 (0.19–1.03)	All	Age, blood pressure, smoking, total cholesterol, education
0.47 (0.20–1.12)	All	Age, blood pressure, smoking, total cholesterol body mass index
0.21 (0.00–0.51)	Randomly selected	Age only
0.48 (0.00–1.00)	Elevated lipids	Age only
0.42 (0.13–1.10)	No history of heart disease	Age only

Table 6-7 Possible rules for choosing from among several estimates of effect for studies where more than one is presented

Choose the estimate adjusted only for age

Choose the estimate adjusted for age and a specified set of confounders that are widely agreed to be true confounders

Choose the "most adjusted" estimate; that is, the estimate with the largest number of variables in the model

Choose the estimate presented in the abstract

of the 95% confidence intervals to assign a weight to the study (see Chapter 7). Studies with wider confidence intervals are weighted less, and when the model is overfitted, the study will be weighted too little in the analysis. The same problem occurs when intermediate variables are included in the multivariate model. These problems should be taken into account when making rules about the choice of estimates to use in the meta-analysis. The model with the most variables may not be the best model.

> *EXAMPLE:* It is believed that part of the reason why estrogen use prevents coronary heart disease is because it lowers total cholesterol and LDL cholesterol and raises HDL cholesterol. Cholesterol is an intermediate variable. Table 6-6 shows that including total cholesterol or LDL cholesterol or HDL cholesterol in the multivariate model aimed at estimating the effect of

estrogen on the risk of cardiovascular disease changes the estimated relative risk of disease in estrogen users from 0.37 to 0.44. The 95% confidence limit is wider when cholesterol is included in the model. Use of an estimate of relative risk from a model with cholesterol in a meta-analysis weights the study less than it would be if the estimate from the model without cholesterol is used.

6.10   INCORPORATING INFORMATION ON STUDY QUALITY

6.10.1   Overview

An estimate of the effect of treatment in a clinical trial or the magnitude of association in a nonexperimental study is valid if it measures what it was intended to measure. Studies of poor quality may yield information that is not valid. Including studies with invalid information in a meta-analysis may render the conclusion of the meta-analysis invalid. For example, a study that calls itself a randomized trial but permits the investigator to tamper with the randomization by placing sicker patients selectively in the treatment arm will yield an estimate of the effect of treatment that is not valid. Including such a study in a meta-analysis would bias the summary estimate of the effect of treatment, and may lead to an invalid conclusion about the effect of the treatment. Similarly, a case-control study of cancer that includes patients whose diagnosis is not confirmed and who may not truly have had cancer will bias the estimate of the association of exposure with disease and may lead to an invalid conclusion about the association of exposure with disease. Studies that allow the investigator to tamper with the randomization or that enroll patients whose diagnosis is not confirmed are of lower quality than studies that protect against tampering and ensure the accuracy of diagnosis of disease. Taking the quality of studies, and implicitly their validity, into account in a meta-analysis has the potential to enhance the validity of the meta-analysis.

Historically, information on study quality has been assessed and incorporated into meta-analysis as follows. First, criteria that measure study quality have been defined, and a scoring system to weight the criteria has been developed. Second, a form for assessing the study in terms of the criteria has been developed. Third, information about the criteria has been abstracted for each eligible study, and the study's quality score has been calculated. Last, the information on study quality has been used to interpret the results of the meta-analysis, most often by stratifying the studies on the basis of their quality scores. Sections 6.10.2 to 6.10.4 describe how quality rating systems have been developed and used historically. Section 6.10.5 discusses the limitations of these historical efforts and Section 6.10.6 describes how quality rating might be improved by greater use of psychometric principles of scale construction.

6.10.2   Developing the Rating System

The development of a quality rating system begins with a listing of elements that define good or poor quality in the study, or in the report on the study. This listing may be based on a published set of criteria defining quality or it may consist of items that a group of experts believe measure quality.

EXAMPLE: Table 6-8 is a list of items that Chalmers et al. (1981) developed to assess the quality of reports of randomized trials.

After the items defining quality are agreed upon, they are translated into a data abstraction form. This form should have the features of good data abstraction forms that are described in detail in Chapter 5.

Next, a scoring system for the quality items is developed. The total score for the "perfect" or highest quality study is defined, and a numerical weight, reflecting the contribution of that item to quality, is assigned to each item. The choice of weights has historically reflected the judgment of the developer of the scale.

EXAMPLE: Table 6-8 also shows the number of points given to various quality items in the system developed by Chalmers et al. (1981) for rating the quality of randomized trials. The total score for a perfect trial is 100 points using this scoring system. The weight of items measuring study design is 0.60. For the items assessing statistical analysis, the weight is 0.30. For items on the presentation of data, it is 0.10.

6.10.3   Abstracting Information About Elements of Quality

The overall goals of abstraction of the information used to rate study quality are reliability and freedom from bias. The main ways to achieve these aims are careful construction of the data collection form, training of the abstractors, and blinding of the abstractors to authors and sources during the entire abstraction process and to the results when abstracting information on quality. It may be desirable to abstract information on different key parts of the study report onto different forms, which are completed in a fixed sequence with each section being filled independent of knowledge of the other sections (Chalmers et al. 1981).

6.10.4   Using the Quality Rating System

Rating study quality implicitly assumes that studies of higher quality scores are more likely to yield valid information than studies with lower quality scores. At the extreme, a study that is excluded from the meta-analysis has a weight of zero and thus contributes no information to the meta-analysis (Laird and Mosteller 1990). The most obvious way to use the quality scores is to stratify the studies by quality score and examine the summary estimate of effect for the studies in each stratum of quality. If these estimates differ, the estimate from the studies with higher quality scores is taken to be the valid one.

EXAMPLE: Steinberg et al. (1991) did a meta-analysis examining the risk of breast cancer in users of estrogen replacement therapy. Table 6-9 shows the mean proportional increase in the risk of breast cancer for each year of use of estrogen for case-control studies in three strata of quality

Table 6-8 Items used to define the quality of reports of randomized trials and points given to each item

Item	Possible Points
Study design	
Description of selection of subject was adequate	3
Description of patients screened was provided	3
Withdrawals and reason for withdrawal were described	3
Therapeutic regimen was defined	3
Appearance of placebo and active drug was identical	1.5
Taste of placebo and active drug was identical	1.5
Randomization was blinded	10
Patients were blinded to treatment group	8
Physicians were blinded to treatment group	8
Physicians and patients were blinded to outcome	4
Number of subjects needed in trial was estimated a priori	3
Adequacy of randomization was evaluated	3
Adequacy of blinding was evaluated	3
Compliance with treatment was assessed	3
Measure of biological activity of the active therapy was made	3
Total	60
Analysis	
Test statistic and probability value were stated	3
For negative trials, statistical power of observed difference was estimated	3
Estimate of confidence interval for effect estimate was given	2
Life table or time series analysis was provided	2
If indicated, regression analysis was done	2
Statistical analysis was appropriate	4
Withdrawals were handled appropriately	4
Side effects were described and statistical analysis of them was done	3
Analysis of subgroups was appropriate	2
Statistician was blinded to treatment group	2
Multiple looks at preliminary results were accounted for	3
Total	30
Presentation	
Starting and stopping dates of accession were provided	2
Analysis to assess baseline comparability of groups was done	2
All events used as endpoints were tabulated	2
Survival curves or data sufficient to construct survival curves were provided	4
Total	10

Source: Chalmers et al. (1981).

Table 6-9 Estimates of the proportional increase in the relative risk of breast cancer per year of estrogen use by quality rating of study for 15 case-control studies eligible for meta-analysis

Reference	Quality Score	Mean Proportional Increase in Risk (95% Confidence Interval)
High-quality score: 71–83		
Wingo et al. (1987)	83	
Bergkvist et al. (1989)	82	
Ross et al. (1980)	75	
Hoover et al. (1981)	72	
Hiatt et al. (1984)	71	
Summary estimate of proportional increase		0.040 (0.030–0.050)
Moderate-quality score: 40–57		
Nomura et al. (1986)	57	
Brinton et al. (1986)	51	
Kaufman et al. (1984)	45	
LeVecchia et al. (1986)	43	
Kelsey et al. (1981)	40	
Summary estimate of proportional increase		-0.008 (-0.002–0.000)
Low-quality score: 15–38		
Hulka et al. (1982)	38	
Jick et al. (1988)	38	
Ravnihar et al. (1979)	26	
Wynder et al. (1978)	25	
Sartwell et al. (1977)	15	
Summary estimate of proportional increase		0.006 (0.000–0.012)

Source: Steinberg et al. (1991); table references cited there.

score—high, moderate, and low. There was a statistically significant association of breast cancer risk with increasing years of estrogen use in studies with high quality scores, but no association or a negative association in studies with moderate and low scores. The authors interpreted these findings as suggestive of a true association of long duration of estrogen use with increased risk of breast cancer.

The quality scores can, in theory, be incorporated into the statistical weights assigned to each study. In this instance, studies that would be excluded can be thought of as having a weight of zero in the analysis (Laird and Mosteller 1990). If low weights are assigned to studies that yield biased estimates of effect, the contribution of their bias to overall score is reduced, although this method does adjust for bias (Laird and Mosteller 1990). Although appealing in theory, the use of quality scores to statistically adjust the results of meta-analysis has not been applied in practice.

6.10.5 Limitations

Rating the quality of studies in a meta-analysis seems desirable on commonsense grounds. It has been shown to be useful in practice. However, the methods for

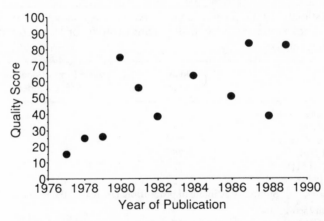

Figure 6-2 Average quality score of case-control studies of estrogen replacement therapy and breast cancer by year of publication. Data are from Steinberg et al. (1991).

developing and using quality rating systems are as yet poorly developed. Even published examples of quality rating of studies for meta-analysis have used ad hoc methods for choosing the items that make up the scale and the weights for each item. The reliability and validity of the quality rating scales have not been evaluated formally.

Quality rating is generally based on the report of the study, and the report of the study may not be an accurate measure of the truth about some elements of quality. For example, if a study report does not state that the patients were blinded to treatment, it does not necessarily mean that the study was unblinded. The study will likely be rated as lower in quality than a study in which the report is explicit about whether or not patients were blinded. A study may suffer from bias, but evidence of the bias may not be detectable in the study report.

The standards for reporting details of the methods of studies seem to have become more stringent over the last decade, and studies published more recently will tend to be rated higher in quality for this reason. It is not certain that the later studies are, in reality, of higher quality.

EXAMPLE: Figure 6-2 shows the quality score of each of the case-control studies included in Steinberg et al.'s (1991) meta-analysis of breast cancer in users of estrogen replacement by year of publication. There is a trend of increasing quality with more recent publication.

It may be impossible to develop scoring systems that will rate studies of different designs fairly in relation to one another. A cohort study is not always superior to a case-control study. Even for experimental studies, the elements of quality may be different for different topics.

It is inherently difficult to rate different sources of bias in relation to one another. For example, is recall bias worse than diagnostic bias? How much worse is information bias than uncontrolled confounding?

6.10.6   Ways to Improve Quality Rating Scales

It would be possible to improve on the ad hoc nature of the procedures to rate quality in several ways. First, a formally chosen panel of experts could be used to select the items to be included in the quality scale. The weights given to each item could also be developed formally by a survey of a small group of experts in the field. Second, the test–retest reliability of the process of scale development could be evaluated by replicating it with another group of experts. Third, the reliability of the abstraction of information needed to complete the scale could be evaluated formally.

The most important advances in development of scales to rate quality might come from application of psychometric principles to development of these scales. Some simple improvements might include asking several observers to rank the studies in relation to each other on an interval scale. The development of systems to rate the quality of studies for meta-analysis is in its infancy, and the general usefulness of quality rating remains to be proven.

7

Statistical Methods in Meta-Analysis

The first goals of the statistical analysis of data from several studies are to estimate a summary measure of effect size, the variance of the summary estimate of effect size, and a confidence interval. The next goal is to derive a summary statistic that can be used for hypothesis testing. Last, statistical methods are used to test the hypothesis that the effects are homogeneous. This chapter describes the statistical methods that address these three goals. It presents methods applicable to comparative studies of the occurrence or nonoccurrence of a dichotomous outcome in relation to either treatment or exposure, also defined as a dichotomy. Statistical methods for meta-analysis of data measured on a continuous scale are described in Chapter 8. Chapter 8 also discusses statistical approaches to publication bias and other statistical issues in meta-analysis.

Section 7.1 discusses the general problem of model choice. Section 7.2 discusses the choice of a measure of effect. Section 7.3 describes the Mantel-Haenszel method as a method for summarizing data. Section 7.4 describes the Peto method. Section 7.5 describes a general variance-based method for summarizing data to estimate rate differences and rate ratios. Section 7.6 describes a variance-based method that relies on information on estimates of the confidence interval to estimate the variance of individual studies. Section 7.7 discusses statistical tests of homogeneity. Section 7.8 describes the DerSimonian-Laird method.

7.1 MODEL CHOICE

7.1.1 Description of the Problem

In an analysis based on a fixed-effects model, inference is conditional on the studies actually done. In an analysis based on a random-effects model, inference is based on the assumption that the studies are a random sample of some hypothetical population of studies. The statistical methods used to combine study results when fixed effects are assumed differ from the methods used when random effects are assumed. The Mantel-Haenszel method (Mantel and Haenszel 1959), the Peto method (Yusuf et al. 1985B), general variance-based methods (Wolf 1986), and the confidence interval methods described independently by Prentice and Thomas (1986) and Greenland (1987) are all methods based on the assumption of fixed effects. The methods described by DerSimonian and Laird (1986) are based on the assumption of random effects.

The question of the appropriate model for a meta-analysis is not just theoretical. The choice of a method based on a fixed-effects or a random-effects model can have important consequences for conclusions based on the meta-analysis.

EXAMPLE: Berlin et al. (1989) compared the results of 22 meta-analyses using the Peto method (Yusuf et al. 1985B), which is an analysis based on the assumption of fixed effects, with that using the DerSimonian-Laird method (DerSimonian and Laird 1986), which is based on the assumption of random effects. One of the meta-analyses they compared was a meta-analysis of randomized trials of interventions using nicotine gum or no gum as an adjunct to smoking cessation in patients seen by primary care physicians (Lam et al. 1987). An analysis of the eligible studies using the Peto method yielded a highly significant ($z = 2.83$, $p < 0.01$) association of treatment with better outcome, whereas the DerSimonian-Laird method yielded a nonsignificant result ($z = 1.58$, $p > 0.05$).

There are strong opinions about the appropriateness of both models. Several statisticians have expressed a preference for the fixed-effects approach (Demets 1987; Peto 1987; Thompson and Pocock 1991); others favor the random-effects approach (Meier 1987; Fleiss and Gross 1991). Examples of results that seem counterintuitive for one method but not the other can be developed for both the fixed-effects model and the random-effects model.

EXAMPLE: Table 7-1 describes two hypothetical randomized clinical trials of the same treatment. In the first trial, 500 patients are treated with the active drug and 500 are given a placebo. The rate of cure in treated patients is 50%. In the patients given placebo, it is 10%. This difference is highly statistically significant ($p < 0.01$). The odds ratio for cure in treated patients is 9.0 (95% C.I. 6.40–12.65). In the second trial, 1000 patients are treated with the active drug, and 1000 are given placebo. The rate of cure in treated patients is 15%. In the control patients, it is 10%. This difference

Table 7-1   Hypothetical data showing how an analysis based on a random-effects model, but not one based on a fixed-effects model, gives a counterintuitive result

Hypothetical Study	Group		Odds of Cure in Treated Patients (95% Confidence Interval)	p
	Active Treatment Cured/N of Subjects	Placebo Cured/N of Subjects		
1	250/500	50/500	9.0 (6.40–12.65)	<0.01
2	150/1000	100/1000	1.59 (1.21–2.08)	<0.01
Summary estimate of odds of cure: fixed-effects model 3.27 (2.67–4.01); $p < 0.01$				
Summary estimate of odds of cure: random-effects model 3.73 (0.67–20.89); $p > 0.05$				

Table 7-2   Hypothetical data showing how an analysis based on the fixed-effects model, but not one based on a random-effects model, gives a counterintuitive result

Hypothetical Study	Odds Ratio	Variance of Logarithm Odds Ratio
Set 1		
1	1.0	0.01
2	6.0	0.01
Summary estimate of odds ratio:	fixed-effects model 2.45 (2.13–2.81)	
95% C.I.	random-effects model 2.45 (0.43–13.75)	
Set 2		
1	2.0	0.01
2	3.0	0.01
Summary estimate of odds ratio:	fixed-effects model 2.45 (2.13–2.81)	
95% C.I.	random-effects model 2.45 (1.65–3.60)	

Source: Fleiss and Gross (1991)

is also highly statistically significant ($p < 0.01$). The odds ratio for cure in treated patients is 1.59 (95% C.I. 1.21–2.08).

The pooled estimate of the odds ratio for cure in treated patients using the method of DerSimonian and Laird (1986), which is based on a random-effects model, is 3.73 (95% C.I. 0.58–24.07). The 95% confidence limit for the estimated effect of treatment includes 1.0. The conclusion based on the results of the meta-analysis is that treatment is not effective. This result is counterintuitive, because one would expect that pooling these two highly statistically significant results should lead to a conclusion that the treatment *is* effective. An analysis using the Mantel-Haenszel method, which is based on a fixed-effects model, gives an estimate of the odds ratio for cure of 3.27 (95% C.I. 2.67–4.01). The conclusion based on this analysis is that treatment is effective, as one would expect intuitively.

Fleiss and Gross (1991) give an example in which an analysis based on a fixed-effects model, but not an analysis based on a random-effects model, yields a result that is counterintuitive. Table 7-2 shows two hypothetical sets of studies. Each set has two studies. In the first set of studies, the odds ratios are 1.0 and 6.0. In the second set of studies, the odds ratios are 2.0 and 3.0. If the variance of the logarithm of the odds ratios for all four studies is the same and equal to 0.01, the pooled odds ratio based on a Mantel-Haenszel analysis, which is based on a fixed-effects model, is 2.45 for both sets of studies. The confidence intervals are also identical (2.13–2.81). These findings are counterintuitive, because one would expect that the confidence interval for the first set of studies would be wider than for the second set of studies because the estimates of the odds ratio are more different. An analysis based on the method of DerSimonian and Laird (1986), which is based on a random-effects model, also yields an odds ratio of 2.45 for both sets of studies. The upper and lower bounds of the confidence intervals are 0.43 and 13.75 for the first set of studies and 1.65 and 3.60 for the second set of studies, which is in line with intuition about the amount of difference in the odds ratios.

7.1.2   Arguments Pertinent to the Controversy

7.1.2.1   Overview

It is generally agreed that the questions addressed by analysis based on the fixed-effects model and based on the random-effects model are different (Bailey 1987). It has also been shown that differences in the results of meta-analysis based on fixed-effects and random-effects models arise only when the study results are not homogeneous (Berlin et al. 1989).

7.1.2.2   Questions Addressed

The random-effects assumption means that the analysis addresses the question, "Will the treatment produce benefit 'on average'?" whereas the fixed-effects assumption means that the analysis addresses the question, "Did the treatment produce benefit on average in the studies at hand?" (Bailey 1987). The random-effects model is appropriate if the question is whether the treatment, or the risk factor, *will* have an effect. If the question is whether the treatment *has* caused an effect *in the studies that have been done,* then the fixed-effects model is appropriate.

Peto (1987) states that analysis using the random-effects model is "wrong" because it answers a question that is "abstruse and uninteresting." Thompson and Pocock (1991) describe as "peculiar" the premise of the random-effects model that studies are representative of some hypothetical population of studies. In contrast, Fleiss and Gross (1991) believe that question addressed by the fixed-effects model is less important than the one addressed by the random-effects model. Consideration of the differences in the questions addressed does not seem to resolve the question of which model to use.

7.1.2.3   Homogeneity

In all the methods based on the assumption of fixed effect, the variance component of the meta-analytic estimate of pooled effect size is composed only of terms for the within-study variance of each component study. The assumption of the random-effects model that studies are a random sample from some population of studies makes it necessary to include a between-study as well as a within-study component of variation in estimation of effect size and statistical significance (Demets 1987; Meier, 1987). Because the random-effects model incorporates a between-study component of variance, it will be more "conservative" than an analysis of the same data based on a method that assumes fixed effects. That is, an analysis based on a random-effects model will generally yield a wider confidence interval and will be less likely to declare a difference significant than an analysis based on fixed effects.

> *EXAMPLES:* In a comparison of the results of meta-analyses using a method based the assumption of fixed effects (the Peto method) and a method based on the assumption random effects (the DerSimonian-Laird method) that was described earlier, Berlin et al. (1989) examined the results of 22 previously published meta-analyses of data from randomized trials. The two methods yielded the same conclusion about the statistical significance of the treatment compared with the control for 19 of the 22 meta-analyses, as shown in Table 7-3. For 3 meta-analyses, the method based on the fixed-effects model was statistically significant, whereas the method based on the random-effects model was not statistically significant. There were no cases where the random-effects analysis yielded a statistically significant result and the fixed-effects analysis did not.

As the between-study variance becomes large (i.e., when there is heterogeneity), the between-study variance term will dominate the weights assigned to the study using the random-effects model, and large and small studies will tend to be weighted equally. In this situation, the results of an analysis based on the fixed-effects model, which weights studies according to their sample size, and the random-effects model may differ considerably, as shown in the hypothetical example given above. When there is not much heterogeneity, the fixed-effects and the random-effects models will both weight studies according to sample size and they will yield results that are essentially identical.

> *EXAMPLE:* In the Berlin et al. (1989) comparison of the Peto method, based on a fixed-effects model, and the DerSimonian-Laird method, based on a random-effects model, there were 14 meta-analyses for which there was no statistical evidence of lack of homogeneity. For these 14 meta-analyses, Table 7-4 shows that there was complete agreement in the conclusions about statistical significance for the two methods.

7.1.3 Summary

There is no empirical basis for preferring the fixed-effects model over the random-effects model or vice versa. Both Greenland and Salvan (1990) and Thompson and Pocock (1991) point out that the choice of a fixed-effects model and a ran-

Table 7-3 Number of statistically significant results for 22 published meta-analyses analyzed using the Peto method (fixed-effects model) and the DerSimonian-Laird method (random-effects model)

Random-Effects Model	Fixed-Effects Model	
	Significant	Not Significant
Significant	8	0
Not significant	3	11

Source: Berlin et al. (1989).

Table 7-4 Number of statistically significant results for 14 published meta-analyses without statistical evidence of heterogeneity analyzed using the Peto method (fixed-effects model) and the DerSimonian-Laird method (random-effects model)

Random-Effects Model	Fixed-Effects Model	
	Significant	Not Significant
Significant	5	0
Not significant	0	9

Source: Berlin et al. (1989).

dom-effects model is secondary to the examination of the reasons for lack of homogeneity. If studies are homogeneous, then the choice between the fixed-effects model and the random-effects model is unimportant, as the models will yield results that are identical. Greenland and Salvan (1990) argue that one shouldn't pool the disparate study results at all. Rather, the results should be reported and modeled, or the reasons for the lack of homogeneity should be examined (Thompson and Pocock 1991).

In this chapter, methods that allow comparative data to be analyzed using the random-effects model as well as methods based on fixed effects are presented. However, use of the random-effects model is not considered to be a defensible solution to the problem of heterogeneity. The random-effects model is generally "conservative." That is, in most situations, use of the model will lead to wider confidence limits and a lower chance of calling a difference "statistically significant." The desire to be conservative is a reason to use the random-effects model, but only in the absence of heterogeneity.

Finally, and perhaps most importantly, the admonition by Jenicek (1989) not to use meta-analysis exclusively to arrive at an average or "typical" value for effect size must be kept in mind. A view of meta-analysis not as a statistical method but as a multicomponent approach for making sense of information focuses attention away from the seemingly insoluble problem of model choice.

Table 7-5 Methods that can be used in meta-analysis according to the underlying model assumption and the type of effect measure for which the method is appropriate

Model Assumption	Methods	Effect Measures
Fixed effects	Mantel-Haenszel	Ratio (typically odds ratio; can be applied to rate ratio and risk ratio)
	Peto	Ratio (approximates the odds ratio)
	General variance-based	Ratio (all types) and difference
Random effects	DerSimonian-Laird	Ratio (all types) and difference

7.2 CHOICE OF EFFECT MEASURE

In randomized trials and cohort studies, the effect of treatment can be estimated as a difference in the rates of disease between the treatment (or exposed) group and the control (or unexposed) group, as the ratio of disease rates measured as incidence density (i.e., with denominators of person-time), as the ratio of rates measured as cumulative incidence (i.e., with denominators of person), or as an odds ratio. In population-based case-control studies, effect can be estimated as an odds ratio or as an attributable risk. It is usual to measure effect using either the odds ratio or a rate ratio in nonexperimental studies. Rate differences are more often used to measure effect in randomized trials. The advantages and disadvantages of ratio and difference measures of effect are discussed by Rothman (1986). These advantages and disadvantages should be considered carefully before choosing a measure of effect for the meta-analysis when more than one effect measure is possible.

The analytic method and the specific formulas used in a meta-analysis are affected by the choice of effect measure. Table 7-5 summarizes the methods that can be used to pool data according to the model (fixed effects or random effects) and the choice of effect measure. The methods are described in detail in subsequent sections.

7.3 MANTEL-HAENSZEL METHOD

7.3.1 Overview

The Mantel-Haenszel method (Mantel and Haenszel 1959) is a well-known method for pooling data across strata. Since studies identified for a meta-analysis are strata, the Mantel-Haenszel method is an appropriate method for analyzing data for a meta-analysis. The method is based on the assumption of fixed effect. It can be used when the measure of effect is a ratio measure, typically an odds ratio. A summary chi-square statistic (Mantel 1963), the variance of the summary effect measure (Robins, Greenland, Breslow 1986), and a test

Table 7-6 Arrangement of data and table notation for application of Mantel-Haenszel and Peto methods

	Exposed[a]	Not Exposed[b]	Total
Diseased	a_i	b_i	g_i
Not diseased	c_i	d_i	h_i
Total	e_i	f_i	n_i

[a]Or treated.
[b]Or not treated.

Table 7-7 Mantel-Haenszel method: Formulas to estimate the summary odds ratio and 95% confidence interval for the summary odds ratio

Summary odds ratio

$$OR_{mh} = \frac{\text{sum(weight}_i \times OR_i)}{\text{sum weight}_i}$$

$$OR_i = \frac{(a_i \times d_i)}{(b_i \times c_i)}$$

$$\text{weight}_i = \frac{1}{\text{variance}_i}$$

$$\text{variance}_i = \frac{n_i}{(b_i \times c_i)}$$

95% confidence interval

$$95\% \text{ C.I.} = e^{\ln OR_{mh} \pm 1.96\sqrt{\text{variance } OR_{mh}}}$$

where variance OR_{mh} is calculated as shown in the appendix to this chapter using the method of Robins, Greenland, and Breslow (1986)

for homogeneity of effect size across strata (Mantel, Brown, Byar 1977) have all been described.

If the data from a study are arranged as shown in Table 7-6 with table notation as shown, Table 7-7 presents the formulas for computing a summary odds ratio using the Mantel-Haenszel method. Kleinbaum, Kupper, and Morgenstern (1982) give formulas that would allow the Mantel-Haenszel method to be applied when data are of the incidence density or cumulative incidence type and the summary estimate of effect is a rate ratio or a risk ratio.

7.3.2 Application of the Mantel-Haenszel Method

EXAMPLE: Table 7-8 shows data from two of the case-control studies of passive smoking and lung cancer that the United States Environmental Protection Agency (1991) identified for the meta-analysis that was discussed in Chapter 2. Both studies were confined to women. In both studies, cases

Table 7-8 Data from two matched case-control studies of lung cancer and passive smoking in women

	Exposed	Not Exposed	Totals
Study 1. Garfinkel, Auerbach, and Joubert (1985)			
Cases	90	44	134
Controls	245	157	402
Totals	335	201	536
Odds ratio = 1.31 95% C.I. (0.85–2.02)			
Study 2. Lam et al. (1987)			
Cases	115	84	199
Controls	152	183	335
Totals	267	267	534
Odds ratio = 1.65 95% C.I. (1.14–2.39)			

and controls were matched on age. The odds ratio for lung cancer in women exposed to passive smoking was 1.31 in the first study (Garfinkel, Auerbach, Joubert 1985) and 1.65 in the second (Lam et al. 1987). The Mantel-Haenszel method is used to estimate a summary odds ratio and its 95% confidence interval for these data as follows:

1. Estimate the variance of the odds ratios for each study where

$$\text{variance}_i = \frac{n_i}{(b_i \times c_i)}$$

Study 1: $\text{variance}_1 = \dfrac{536}{(44 \times 245)} = 0.050$

Study 2: $\text{variance}_2 = \dfrac{534}{(84 \times 152)} = 0.042$

2. Calculate the weights for each study where

$$\text{weight}_i = \frac{1}{\text{variance}_i}$$

Study 1: $\text{weight}_1 = \dfrac{1}{\text{variance}_1} = \dfrac{1}{0.050} = 20.00$

Study 2: $\text{weight}_2 = \dfrac{1}{\text{variance}_2} = \dfrac{1}{0.042} = 23.81$

3. Calculate the product of the weights and the ORs:

Study 1: $\text{product}_1 = \text{odds ratio}_1 \times \text{weight}_1 = 20.00 \times 1.31 = 26.20$

Study 2: $\text{product}_2 = \text{odds ratio}_2 \times \text{weight}_2 = 23.81 \times 1.65 = 39.29$

4. Calculate the sum of the weights:

   sum of weights $= 20.00 + 23.81 = 43.81$

5. Calculate the sum of the product of the weights and the ORs:

   sum of products $= 26.20 + 39.29 = 65.49$

6. Estimate the OR_{mh} by dividing the sum of the products by sum of the weights:

   summary $OR_{mh} = \dfrac{65.49}{43.81} = 1.49$

7. Estimate the variance of OR_{mh} using the method of Robins, Greenland, and Breslow (1986) using the formulas in the appendix to this chapter:

   variance $OR_{mh} = 0.019$

8. Estimate the 95% confidence interval where

   $$CI = e^{\ln OR + (1.96 \times \sqrt{\text{variance OR mh}})}$$

   upper bound $= e^{0.399 + (1.96 \times \sqrt{0.019})} = e^{0.399 + 0.270} = e^{0.669} = 1.95$
   lower bound $= e^{0.399 - (1.96 \times \sqrt{0.019})} = e^{0.399 + 0.270} = e^{0.129} = 1.14$

7.3.3 Strengths and Limitations

A number of factors argue for the use of the Mantel-Haenszel method when it is applicable and the data are available. First, the test based on the Mantel-Haenszel chi-square has optimal statistical properties, being the uniformly most powerful test (Radhakrishna 1965). Second, the Mantel-Haenszel estimate of effect equals one only when the Mantel-Haenszel chi-square is equal to zero, which provides a mathematical connection of the effect estimate with the summary statistic. Third, a number of widely available computer programs (e.g., EPIINFO, EGRET, STATXACT) include programs to apply the method.

There are some problems with the Mantel-Haenszel approach that limit its usefulness in practice. First, application requires that data to complete a 2×2 table of outcome by treatment (or exposure by disease) for each study be available. If data that would allow construction of a 2×2 table for a study are unavailable, the study must be excluded. Exclusion has the potential to result in bias.

More important, the Mantel-Haenszel approach ignores confounding that is not taken into account in the design of the study. In randomized trials and in case-control studies that have been matched for confounders, as in the example, the failure of the method to control for confounding is not a problem. In a meta-analysis where some studies are case-control studies that are unmatched for age, or only coarsely matched for age, age differences between cases and controls will not be taken into account in the summary estimate of effect using this method. Highly misleading conclusions can result in this situation and in any other situation where the result of an analysis controlling for confounding is different from the crude result.

Table 7-9   Peto method: Formulas to estimate the summary odds ratio and 95% confidence interval for the summary odds ratio

---

Summary odds ratio

$$OR_p = e^{\text{sum}(O_i - E_i)/\text{sum variance}_i}$$

or

$$\ln OR_p = \frac{\text{sum}(O_i - E_i)}{\text{sum variance}_i}$$

$$E_i = \frac{(e_i \times g_i)}{n_i}$$

$$\text{variance}_i = \frac{(E_i \times f_i \times h_i)}{(n_i \times (n_i - 1)}$$

95% confidence interval

$$95\% \text{ C.I.} = e^{\ln OR_p \pm 1.96/\sqrt{\text{sum variance}_i}}$$

where the variance$_i$ are calculated as shown above

7.4   PETO METHOD

7.4.1   Overview

The Peto method (Yusuf et al. 1985B) is a modification of the Mantel-Haenszel method. Like the Mantel-Haenszel method, it is based on a fixed-effects model. It is an alternative method for pooling data when the effect measure of interest is a ratio measure. Like the Mantel-Haenszel method, it is computationally simple. It has been used frequently in meta-analyses of randomized trials.

Table 7-9 shows the formulas for using the Peto method for combining data to estimate a summary measure of effect, which will be called the Peto odds ratio (OR_p). The formulas are based on an arrangement of data and table notation that was shown in Table 7-6.

7.4.2   Application of the Peto Method

EXAMPLE   Table 7-10 presents data from the two largest randomized trials of antiplatelet treatment for patients with a transient ischemic attack or ischemic stroke identified by the Antiplatelet Trialists' Collaboration (1988) as eligible for their meta-analysis. In the first study, 14.6% of the patients treated with active drug had an important vascular event (first myocardial infarction, stroke, or vascular death) compared with 21.1% in the control group. In the second study, 22.1% of patients in the active treatment group had an important vascular event compared with 25.1% in the control group. The Peto method is used to estimate a summary odds ratio and its 95% confidence interval for these data as follows:

Table 7-10 Data from two randomized trials of antiplatelet therapy for treatment of transient ischemic attack of stroke

	Treatment	Control	Total
Study 1. European Stroke Prevention Study Group (1987)			
Events[a]	182	264	446
Nonevents	1,068	986	2,054
Total	1,250	1,250	2,500
Odds ratio = 0.64			
Study 2. United Kingdom Transient Ischemic Attack Aspirin Trial (unpublished)			
Events[a]	348	204	552
Nonevents	1,273	610	1,883
Total	1,621	814	2,434
Odds ratio = 0.82			

[a]Important vascular events (first myocardial infarction, stroke, vascular death).

Source: Antiplatelet Trialists' Collaboration (1988); table references cited there.

1. Calculate the expected number of events in the treatment group for each study where

$$E_i = \frac{(e_i \times g_i)}{n_i}$$

Study 1: $E_1 = \frac{(1250 \times 446)}{2500} = 223.0$

Study 2: $E_2 = \frac{(1621 \times 552)}{2435} = 367.5$

2. Calculate the difference in the observed and expected number of events in the treatment group for each study:

Study 1: $O_1 - E_1 = 182 - 223.0 = -41.0$
Study 2: $O_2 - E_2 = 348 - 367.5 = -19.5$

3. Estimate the variance of the observed minus expected for each study where

$$\text{variance}_i = \frac{(E_i \times f_i \times h_i)}{n_i(n_i - 1)}$$

Study 1: $\text{variance}_1 = \frac{(223.0 \times 1250 \times 2054)}{(2500 \times 2499)} = 91.6$

Study 2: $\text{variance}_2 = \frac{(367.5 \times 814 \times 1883)}{(2435 \times 2434)} = 95.0$

4. Calculate the sum of the values of observed minus expected:

sum $(O_i - E_i) = -41.0 + -19.5 = -60.5$

5. Calculate the sum of the variances:

sum variance$_i$ = 91.6 + 95.0 = 186.6

6. Estimate the natural logarithm of the OR$_p$ by dividing the sum of the sum of observed minus expected (result of step 4) by the sum of variances (result of step 5):

$$\ln OR_p = \frac{-60.5}{186.6} = -0.32$$

8. Estimate the summary odds ratio by taking e to the power of the result of step 6:

$$OR_p = e^{-0.32} = 0.72$$

9. Estimate the 95% confidence interval where

$$CI = e^{OR_p \pm 1.96/\sqrt{\text{sum of variance}_i}}$$

upper bound = $e^{-0.32+0.14=0.84}$ = 0.84
lower bound = $e^{-0.32-0.14=0.84}$ = 0.63

7.4.3 Strengths and Limitations

The Peto method, like the Mantel-Haenszel method, requires data to complete a 2×2 table for every study to be included in the meta-analysis. If data to complete a 2×2 table are not available, the study must be excluded. Exclusion has the potential to cause bias. The Peto method will rarely be useful in the analysis of data from nonexperimental studies because appropriate data from every eligible study are rarely available. Furthermore, like the Mantel-Haenszel method, the Peto method cannot incorporate confounding that is not taken care of by design. Last, Greenland and Salvan (1990) show that the Peto method can yield a biased summary estimate of effect when some of the individual studies are unbalanced. They recommend against using it to analyze data from nonexperimental studies. For nonexperimental studies, the advantages of the Peto method over the Mantel-Haenszel method in terms of computational simplicity probably do not outweigh the disadvantage due to the possibility of bias, especially since it has the same limitations as the Mantel-Haenszel method (Greenland and Salvan 1990). For experimental studies, where lack of balance is less a problem, the method is acceptable, although the Mantel-Haenszel method is also acceptable.

7.5 GENERAL VARIANCE-BASED METHODS

7.5.1 Overview

The Mantel-Haenszel method and the Peto method generally apply to estimation of effects measured on a ratio scale. When the goal of the meta-analysis is to derive a summary estimate of a difference measure, the following general, variance-based method applies:

$$RD_s = \frac{\text{sum}(w_i \times RD_i)}{\text{sum } w_i}$$

where

$$w_i = \frac{1}{\text{variance}_i}$$

Here, RD_s is the summary estimate of the rate difference, w_i is the weight assigned to the ith study, RD_i is the rate difference from the ith study, and variance$_i$ is an estimate of the variance of the rate difference in the ith study (Prentice and Thomas 1986; Wolf 1986; Greenland 1987).

The formulas for estimating the variance of the difference in two rates differ according to whether the data to be pooled are cumulative incidence or incidence density data. These formulas are provided in Kleinbaum, Kupper, and Morgenstern (1982).

A 95% confidence interval for an estimate of effect derived from the preceding general equation can be estimated as follows:

$$95\% \text{ CI} = RD_s + 1.96 \times \sqrt{\text{variance}_s}$$

where

$$\text{variance}_s = \frac{1}{\text{sum weight}_i}$$

When effect size is measured as an incidence density ratio (i.e., as the ratio of two incidence rates measured with person–time in the denominator) or as a risk ratio (i.e., as the ratio of two incidence rates measured as cumulative incidence), the general variance-based method given above can be applied after logarithmic transformation as follows:

$$\ln RR_s = \frac{\text{sum}(w_i \times \ln RR_i)}{\text{sum } w_i}$$

where

$$w_i = \frac{1}{\text{variance } y_i}$$

Formulas to estimate the variance of the incidence density ratio and the risk ratio are also provided by Kleinbaum, Kupper, and Morgenstern (1982).

A 95% confidence interval for the summary ratio is estimated as

$$95\% \text{ C.I.} = e^{RR_s \pm 1.96 \times \sqrt{\text{variance}_s}}$$

where

$$\text{variance}_s = \frac{1}{\text{sum weight}_i}$$

7.5.2 Application of the General Variance-Based Method

7.5.2.1 Rate Difference with Cumulative Incidence Data

Example The data from the two studies of antiplatelet treatment shown in Table 7-10 can be expressed as the difference in the rates of disease in the active treatment group and the control group. In the first study, the difference in the rate of important vascular events between the treated and untreated patients was 6.5 events per 100 (0.065). In the second study, the difference in rates was 3.6 events per 100 (0.036). The general variance-based method can be used to estimate a summary measure of the rate difference and a 95% confidence interval as follows:

1. Estimate the variance of the rate difference for each study where

$$\text{variance RD}_i = \frac{(g_i \times h_i)}{(e_i \times f_i \times n_i)}$$

Study 1: $\text{variance}_1 = \dfrac{(446 \times 2054)}{(1250 \times 1250 \times 2500)} = 0.00023$

Study 2: $\text{variance}_2 = \dfrac{(552 \times 1883)}{(1621 \times 814 \times 2435)} = 0.00032$

2. Estimate weight for each study where

$$\text{weight}_i = \frac{1}{\text{variance}_i}$$

Study 1: $\text{weight}_1 = \dfrac{1}{0.00023} = 4347.8$

Study 2: $\text{weight}_2 = \dfrac{1}{0.00032} = 3125.0$

3. Calculate the sum of the weights:

sum of weights = 4347.8 + 3125.0 = 7472.8

4. Calculate the product of the weights and the RDs:

Study 1: $\text{product}_1 = \text{weight}_1 \times \text{RD}_1 = 4347.8 \times 0.065 = 282.6$
Study 2: $\text{product}_2 = \text{weight}_2 \times \text{RD}_2 = 3125.0 \times 0.036 = 112.5$

5. Calculate the sum of the product of weights and rate differences:

sum of products = 282.6 + 112.5 = 395.1

6. Divide the sum of products (result of step 5) by the sum of weights (result of step 3):

$$\text{RD}_s = \frac{395.1}{7472.8} = 0.053 \qquad \text{or} \quad 5.3 \text{ deaths per } 100$$

7. Estimate the 95% confidence interval where

$$CI = RD_s + 1.96 \times \sqrt{variance_s} \quad \text{and} \quad variance_s = \frac{1}{\text{sum weight}_i}$$

$$\text{lower bound} = 0.053 - \left(1.96 \times \sqrt{\left(\frac{1}{7472.8}\right)}\right) = 0.053 - 0.023 = 0.030$$

$$\text{upper bound} = 0.053 + \left(1.96 \times \sqrt{\left(\frac{1}{7472.8}\right)}\right) = 0.053 + 0.023 = 0.076$$

7.5.2.2 Rate Ratio with Cumulative Incidence Data

EXAMPLE: The data in Table 7-10 from the randomized trials of anti-platelet treatment could have been analyzed to derive an estimate of the ratio of the rates of important vascular disease in the treated and untreated. In the first study, the ratio of the rate of important vascular disease in the treated group compared with the control group is 0.69. In the second study, it is 0.86. The summary measure of the rate ratio and its 95% confidence limit are estimated as follows:

1. Estimate the variance of the rate ratio where

$$variance_i = \frac{(h_i \times n_i)}{(e_i \times f_i \times g_i)}$$

Study 1: $variance_1 = \dfrac{(2054 \times 2500)}{(1250 \times 1250 \times 446)} = 0.0074$

Study 2: $variance_2 = \dfrac{(1883 \times 2435)}{(1621 \times 814 \times 552)} = 0.0063$

2. Calculate the weights for each study where

$$weight_i = \frac{1}{variance_i}$$

Study 1: $weight_1 = \dfrac{1}{variance_1} = \dfrac{1}{0.0074} = 135.1$

Study 2: $weight_2 = \dfrac{1}{variance_2} = \dfrac{1}{0.0063} = 158.7$

3. Calculate the sum of weights:

sum of weights $= 135.1 + 158.7 = 293.8$

4. Calculate the natural logarithm of each rate ratio:

Study 1: $\ln RR_1 = \ln 0.69 = -0.371$
Study 2: $\ln RR_2 = \ln 0.86 = -0.151$

5. Calculate the product of the weights and the natural logarithm the rate ratios:

Study 1: product $_1$ = weight$_1$ × ln RR$_1$ = 135.1 × (−0.371) = −50.12
Study 2: product $_2$ = weight$_2$ × ln RR$_2$ = 158.7 × (−0.151) = −23.96

6. Calculate the sum of the product of weights and the natural logarithm the rate ratios:

   sum of products = −50.12 + (−23.96) = −74.08

7. Calculate natural logarithm of the summary RR where

   $$\ln RR_s = \frac{\text{sum of products}}{\text{sum of weights}}$$
   $$\ln RR_s = \frac{-74.08}{298.8} = -0.248$$

8. Estimate the summary measure of effect by taking e to the power of the result of step 7:

   $$RR_s = e^{-0.248} = 0.78$$

9. Estimate the 95% confidence interval where

   $$CI = e^{\ln \text{ summary } RR \pm 1.96 \times \sqrt{\text{variances}}} \quad \text{and} \quad \text{variance}_s = \frac{1}{\text{sum weight}_i}$$

   lower bound = $e^{-0.248 - 1.96 \times \sqrt{1/293.8}} = e^{-0.362} = 0.70$
   upper bound = $e^{-0.248 + 1.96 \times \sqrt{1/293.8}} = e^{-0.134} = 0.87$

This result can be compared with the result based on the Peto method based on the same data, which yielded an estimated odds ratio 0.72 (95% C.I. 0.63–0.84).

7.5.3 Strengths and Limitations

These general variance-based methods allow meta-analysis with rate difference as the measure of effect. When the measure of effect of interest is an odds ratio, these methods have no obvious advantages over the Mantel-Haenszel method. Estimating the variances and the weights for each study is computationally more burdensome.

The general variance-based approach is the basis for the confidence interval approach, a widely applicable procedure for summarizing data from nonexperimental studies that is described in the following section.

7.6 GENERAL VARIANCE-BASED METHODS THAT USE CONFIDENCE INTERVALS

7.6.1 Overview

The problem with all of the previously described methods is that they require construction of 2 × 2 tables for every study in the meta-analysis and they ignore confounding. Prentice and Thomas (1986) and Greenland (1987) independently described a general variance-based method for meta-analysis where the effect measures are ratio measures that require only information on each study's estimate of relative risk and its 95% confidence interval. The estimate of the 95% confidence interval from each study is used to estimate the variance of each study's effect measure. The following general formulas apply:

$$\ln RR_s = \frac{\text{sum}(w_i \times \ln RR_i)}{\text{sum } w_i}$$

where

$$w_i = \frac{1}{\text{variance } RR_i}$$

The RR_i are estimates of relative risk and may have been measured as odds ratios, rate ratios, or risk ratios.

The formula for estimating variance from the 95% confidence interval given by Prentice and Thomas (1986) is

$$\text{variance } RR_i = \left[\frac{\ln (RR_i/RR_l)}{1.96} \right]^2$$

or, equivalent when the confidence interval is symmetric (i.e., when $RR_i / RR_l = RR_u / RR_i$),

$$\text{variance } RR_i = \left[\frac{\ln (RR_u/RR_i)}{1.96} \right]^2$$

where RR_i is the estimate of the relative risk in the ith study, RR_u is the upper bound of the 95% confidence interval for that study, and RR_l is the lower bound of the 95% confidence interval for that study

A 95% confidence limit for the estimated relative risk is calculated as

$$95\% \text{ CI} = e^{\ln RR_s \pm 1.96 \times \sqrt{\text{variance}_s}}$$

and

$$\text{variance}_s = \frac{1}{\text{sum weight}_i}$$

7.6.2 Application of Confidence Interval Methods

EXAMPLE A summary estimate of the relative risk of lung cancer and environmental tobacco smoke can be estimated using the confidence interval approach for tha data from Table 7-8. The confidence interval approach is used to estimate a summary relative risk and a 95% confidence interval as follows:

1. Take the natural logarithm of the estimated relative risk for each study:

   Study 1: $\ln RR_1 = \ln 1.31 = 0.270$
   Study 2: $\ln RR_2 = \ln 1.65 = 0.501$

2. Estimate the variance of the relative risk for each based on the estimated relative risk for that study and on the lower bound of the 95% confidence interval for that study where

$$\text{variance RR}_i = \left[\frac{\ln (RR_i/RR_l)}{1.96} \right]^2$$

$$\text{Study 1: variance}_1 = \left[\frac{\ln (1.31/0.85)}{1.96} \right]^2 = 0.049$$

$$\text{Study 2: variance}_2 = \left[\frac{\ln (1.65/1.14)}{1.96} \right]^2 = 0.036$$

3. Estimate the weight of each study where

$$\text{weight}_i = \frac{1}{\text{variance}_i}$$

$$\text{Study 1: weight}_1 = \frac{1}{\text{variance}_1} = \frac{1}{0.049} = 20.41$$

$$\text{Study 2: weight}_2 = \frac{1}{\text{variance}_2} = \frac{1}{0.036} = 27.78$$

4. Calculate the sum of the weights:

sum of weights = 20.41 + 27.78 = 48.19

5. Calculate the product of the weight and the natural logarithm of the estimated relative risk:

Study 1: product$_1$ = 20.41 × 0.270 = 5.511
Study 2: product$_2$ = 27.78 × 0.501 = 13.918

6. Calculate the sum of the products:

sum of products = 5.511 + 13.918 = 19.429

7. Estimate the summary measure of effect where

$$RR_s = e^{(\text{sum of products/sum of weight}_s)}$$
$$RR_s = e^{19.429/48.19} = e^{0.403} = 1.50$$

8. Estimate 95% confidence interval for the summary measure of relative risk where

$$\text{C.I.} = e^{\ln RR + (1.96 \times \sqrt{\text{variance}_s})} \quad \text{and} \quad \text{variance}_s = \frac{1}{\text{sum weight}_i}$$

upper bound $e^{0.403+(1.96 \times \sqrt{1/48.19})} = e^{0.685} = 1.98$
lower bound $e^{0.403-(1.96 \times \sqrt{1/48.19})} = e^{0.121} = 1.13$

The summary estimate of the relative risk of lung cancer based on the Mantel-Haenszel method was 1.49 (95% C.I. 1.13–1.98).

7.6.3 Handling Studies with Missing Confidence Intervals

Sometimes a study will not present an estimate of a 95% confidence interval. A confidence interval can occasionally be estimated from the data provided in the study report. It is sometimes worthwhile to contact the investigator to obtain the information that would allow the confidence interval to be estimated if the data to do so are not provided in the study report. Formulas to estimate confidence intervals for a variety of study designs and data presentations are given by Kleinbaum, Kupper, and Morgenstern (1982).

It is important to be sure that the confidence interval estimated from the raw data presented in a paper is correct. Errors in estimation of confidence intervals can lead to *substantial* bias in the summary estimate of relative risk.

EXAMPLE: Data from case-control studies of the risk of ovarian cancer in women who had a family history of ovarian cancer were abstracted from case-control studies of ovarian cancer in order to do a meta-analysis. The 95% confidence intervals for the estimated relative risk of ovarian cancer in women with a family history of ovarian cancer had not been calculated in some of the studies, but the raw data necessary to make these calculations were presented in all of them. Table 7-11 shows the 95% confidence intervals for the studies as they were originally calculated. It also shows the summary estimate of the relative risk of ovarian cancer in women with a family history of ovarian cancer based on the estimates shown in the table. The summary estimate is 6.52 (95% C.I. 5.10–8.33). One study, study 3, had a very high weight in the analysis because the confidence interval was fairly narrow. The fact that the study carried so much weight led to recalculation of the 95% confidence interval for the study, and it was found that a mistake in the original calculation had been made. The summary estimate of relative risk based on the correct intervals, also shown in Table 7-11, is 4.52 (95% C.I. 1.13–3.10).

Table 7-11 Estimated relative risk of ovarian cancer in women with a family history of ovarian cancer based on incorrect calculation of confidence interval for study 3

Study	Estimated Relative Risk	Incorrect (95% Confidence Interval)	Correct (95% Confidence Interval)
1	9.25	(0.49–173.1)	(0.49–173.1)
2	18.20	(4.8–59.0)	(4.8–69.0)
3	11.32	(8.2–18.3)[a]	(0.6–211.3)
4	3.6	(1.8–7.2)	(1.8–7.2)
5	3.3	(1.1–9.4)	(1.1–9.4)
6	1.90	(1.1–3.6)	(1.1–3.6)
Summary relative risk and incorrect C.I.:			6.52 (5.10–8.33)
Summary relative risk and correct C.I.:			4.52 (1.13–3.10)

[a]Incorrect confidence interval.

Greenland (1987) suggests ways to use information on p values to estimate 95% confidence intervals when the p value is the only information available in the study report. These methods are test-based methods and they should be applied recognizing the limitations and problems with test-based methods (Kleinbaum, Kupper, Morgenstern 1982; Greenland 1987). When a p value has been used to estimate the confidence interval for a study, it may be wise to do the meta-analysis with and without the study to determine whether the conclusion is dependent on inclusion, a kind of sensitivity analysis.

7.6.4   Strengths and Limitations

Because the variance estimates are based on the adjusted measure of effect and on the 95% confidence interval for the adjusted measure, the confidence interval methods do not ignore confounding. Since most modern studies present confidence intervals for estimates of relative risk, or they can be estimated from data in a publication when the authors have not done so themselves, few studies are excluded because of missing data.

7.7   STATISTICAL TESTS OF HOMOGENEITY

The problem of lack of homogeneity was discussed in Section 7.2. Table 7-12 gives the formulas that can be used to calculate a statistic to test the hypothesis that the effect sizes are equal in all of the studies. These tests are variously referred to as tests of homogeneity and tests of heterogeneity. They are called tests of homogeneity here.

All the formulas follow the general pattern of testing the sum of the weighted

Table 7-12   Formulas to calculate a statistic for a test of homogeneity of effects

---

Mantel-Haenszel method

$$Q = \text{sum}[\text{weight}_i \times (\ln OR_{mh} - \ln OR_i)^2]$$

where OR_{mh} and the weight_i are estimated as shown in Table 7-7

Peto method

$$Q = \text{sum}[\text{weight}_i \times (O_i - E_i)^2] - \frac{\text{sum}(O_i - E_i)^2}{\text{sum variance}_i}$$

where the O_i, E_i, weight_i, and variance_i are estimated as shown in Table 7-9

General variance-based method

$$Q = \text{sum}[\text{weight}_i \times (\ln OR_s - \ln OR_i)^2]$$

where OR_s and the weight_i are estimated as described in the text

---

Note: Q is referred to the chi-square distribution with degrees of freedom equal to the number of studies minus 1.

difference between the summary effect measure and the measure of effect from each study. The statistic calculated by these formulas is referred to the chi-square distribution, although it usually is called Q. The number of degrees of freedom of Q is equal to the number of studies minus one.

When the p value for the test of homogeneity exceeds some critical value of alpha (usually 0.05), the hypothesis of homogeneity is rejected. In rejecting the hypothesis that the studies are homogeneous, one can conclude that the studies are not measuring an effect of the same size; that is, the studies are heterogeneous.

> *EXAMPLE:* A statistical test of homogeneity for the studies of stroke and estrogen replacement therapy that were described in Table 6-1 was 17.3 with 9 degrees of freedom ($p = 0.044$). At a critical value of 0.05, the hypothesis of homogeneity of effects is rejected.

When there is statistical evidence of lack of homogeneity, calculating a pooled estimate of effect size is of dubious validity. Attempts to explain the lack of homogeneity based on consideration of study design or other characteristics of the studies may be useful (Greenland 1987; Jenicek 1989; Greenland and Salvan 1990; Greenland and Longnecker 1992).

The power of statistical tests of homogeneity is low, and the failure to reject the hypothesis that the studies are homogeneous does not prove that the studies are measuring the same quantity.

7.8 DERSIMONIAN AND LAIRD METHOD

7.8.1 Overview

The DerSimonian and Laird (1986) method is based on the random-effects model. Formulas for applying the DerSimonian and Laird method summarizing studies in the case where effects are measured as odds ratios are given by Fleiss and Gross (1991) as follows:

$$\ln OR_{dl} = \frac{\text{sum}(w_i^* \times \ln OR_i)}{\text{sum } w_i^*}$$

where OR_{dl} is the DerSimonian-Laird summary estimate of the odds ratio, w_i^* is the DerSimonian-Laird weighting factor for the ith study, and OR_i is the odds ratio from the ith study.

The weighting factors, w_i^*, are estimated as

$$w_i^* = \frac{1}{[D + (1 / w_i)]}$$

where

$$w_i = \frac{1}{\text{variance}_i}$$

The variance$_i$ for each study is estimated using the Mantel-Haenszel method described previously and:

$$D = \frac{[Q - (S - 1)] \times \text{sum } w_i}{[(\text{sum } w_i)^2 - \text{sum } (w_i^2)]}$$

where S is the number of studies and

$$Q = \text{sum } w_i \, (\ln \text{OR}_i - \ln \text{OR}_{mh})$$

The OR$_i$ are odds ratios from the ith study, OR$_{mh}$ is the summary odds ratio calculated using the Mantel-Haenszel method, and w_i is the weight of ith study calculated using the Mantel-Haenszel method as described in Section 7.3.

A 95% confidence interval for the DerSimonian-Laird summary estimate of the odds ratio derived from the above equation is estimated as follows:

$$\text{CI} = e^{\ln \text{OR}_{dl} + 1.96 \times \sqrt{\text{variance}_s^*}}$$

where

$$\text{variance}_s^* = \text{sum weight}_i^*$$

7.8.2 Application of the DerSimonian-Laird Method

EXAMPLE The data in Table 7-8 were analyzed using the Der-Simonian-Laird method to obtain a summary odds ratio and a 95% confidence interval using the formulas shown above. The summary odds ratio using the DerSimonian-Laird method is 1.50 (95% C.I. 1.20–1.87). Table 7-13 shows the summary odds ratio for the data in Table 7-8 as estimated using the Mantel-Haenszel method, the confidence limit method, and the method of DerSimonian and Laird. In this example, the estimates do not differ much. This is not surprising, since the number of studies is small and there is no evidence of lack of homogeneity, which is to be expected given the small number of studies.

7.8.3 Strengths and Limitations

If the random-effects model is considered on theoretical grounds to be the appropriate model for meta-analysis, then the method of DerSimonian and Laird (1986) is the appropriate method to use in the analysis. The results of an analysis based on the DerSimonian-Laird method will differ from an analysis based on the fixed-effects model only if there is lack of homogeneity. In this case, combining the studies may be questionable. Thus, in the very situations where application of the method matters, a simple summary estimate of effect is inappropriate. Use of a random-effects model should not substitute for exploration of the reasons for

Table 7-13 For data shown in Table 7-8, summary estimate of relative risk and 95% confidence intervals from three methods of analysis

Method	Model Assumption	Estimated Relative Risk (95% Confidence Interval)
Mantel-Haenszel	Fixed effects	1.49 (1.14–1.95)
Confidence interval	Fixed effects	1.50 (1.13–1.98)
DerSimonian-Laird	Random effects	1.50 (1.20–1.87)

the lack of homogeneity. This method does not correct for bias, failure to control confounding, or for any other cause of lack of homogeneity.

Methods based on the random-effects model tend to give high weight to small studies. Since small studies may reflect publication bias, use of the model may emphasize poor evidence at the expense of good evidence (Thompson and Pocock 1991). The computational burden of doing an analysis using the method of DerSimonian and Laird is greater than the computational burden of using any of the other methods. None of widely available software packages contains a program or macro that allows one to do an analysis using the method.

One approach is to use both models and to test the dependence of the conclusions of the analysis on the model assumption. This is a kind of sensitivity analysis. It is considered further in Chapter 13.

APPENDIX

Formulas to estimate the variance of the Mantel-Haenszel summary odds ratio as described by Robins, Greenland, and Breslow (1986):

$$\text{variance}_{mh} = \left(\frac{\text{sum } F}{2 \text{ sum } R^2}\right) + \left[\frac{\text{sum } G}{(2 \times \text{sum } R \times \text{sum } S)}\right] + \left(\frac{\text{sum } H}{2 \text{ sum } S^2}\right)$$

where

$$F = a_i \times d_i \times \frac{(a_i + d_i)}{n_i^2}$$

$$G = \frac{[a_i \times d_i \times (b_i + c_i)] + (b_i \times c_i \times (a_i + d_i))]}{n_i^2}$$

$$H = \frac{b_i \times c_i \times (b_i + c_i)}{n_2^2}$$

$$R = \frac{a_i \times d_i}{n_i}$$

$$S = \frac{b_i \times c_i}{n_i}$$

The table notation is as follows:

	Exposed	Not Exposed	Total
Diseased	a_i	b_i	
Not diseased	c_i	d_i	
Total			n_i

8

Other Statistical Issues in Meta-Analysis

Studies that measure effects on a continuous scale are often the subject of meta-analysis. The goal of meta-analysis is often not simply to estimate an overall measure of effect, but to estimate the relationship between disease and some measure of intensity of exposure. Statistical models can be used in meta-analysis to examine the reasons for heterogeneity or to explore the relationship between various study descriptions and the effect size. Last, there are some statistical methods that have been used in meta-analysis that are not recommended. These methods, and the reasons why they are not recommended, need to be understood. This chapter addresses all of these topics.

Section 8.1 describes statistical methods for meta-analysis of effects measured on a continuous scale. Section 8.2 presents the statistical methods to derive a summary estimate of the trend of effect with increasing level of exposure. Section 8.3 discusses modeling. Section 8.4 describes some methods for estimating a summary statistic in meta-analysis that are *not* recommended. Section 8.5 describes several statistical approaches to the problem of publication bias and discusses why these approaches are problematic.

8.1 MEASURES ON A CONTINUOUS SCALE

8.1.1 Overview

Blood pressure, hemoglobin concentration, and level of depression are three examples of continuous measures that might be outcome measures in experimental or nonexperimental studies. In a meta-analysis of studies where effect size is measured on a continuous scale, there are two situations to be considered. In the first situation, all of the eligible studies use the same measure of effect. For example, all of the studies may measure the effect of the intervention on blood pressure, serum cholesterol level, or depression as assessed by the Beck depression scale. In the second situation, all of the eligible studies addressed the same question, but the measure of effect was made using different instruments and thus different scales. For example, in a series of studies on the effect of the intervention on depression, some studies might have used the CES-D depression scale and some might have used the Hamilton depression scale. Methods to address the first situation are directly related to analysis of variance, and they are described in textbooks, usually under the heading of weighted studies. Most of the statistical writing about meta-analysis of studies specifically dealing with measures of effect on a continuous scale appears in the social science literature and addresses the second situation (Glass, McGaw, Smith 1981; Hedges 1982; Wolf 1986). Methods to handle both situations are described here.

8.1.2   When Outcome Is Measured on the Same Scale

Cochran (1954) comprehensively described methods to combine results from different experiments. These methods are an extension of analysis of variance where the "groups" are studies. Using the analysis of variance analogy, application of the fixed-effects model to a continuous measure of effect is application of "Model

Table 8-1 Continuous measure of effect with all measures on the same scale: formulas to estimate the summary measure of effect, a statistic to test homogeneity, and 95% confidence interval

---

Summary measure of effect size

$$\text{mean}_s = \frac{\text{sum (weight}_i \times \text{mean}_i)}{\text{sum weight}_i}$$

$$\text{weight}_i = \frac{1}{\text{variance}_i}$$

where the variance$_i$ are the SD_i^2 as calculated using methods described in introductory textbooks of statistics (e.g., Armitage and Berry 1987)

Test of homogeneity

$$Q = \text{sum [weight}_i \times (\text{mean}_s - \text{mean}_i)^2]$$

where the weight$_i$ are estimated as described above; Q is referred to the chi-square distribution with degrees of freedom equal to the number of studies minus 1

95% confidence interval

$$95\% \text{ C.I.} = \text{mean}_s \pm (1.96 \times \sqrt{\text{variance}_s})$$

$$\text{variance}_s = \frac{1}{\text{sum weight}_i}$$

where the weight$_i$ are estimated as described above

Table 8-2 Change in Kutzke Disability Status Scale at two years in four randomized trials of the effect of azathioprine treatment in multiple sclerosis

Study	Treated			Control		
	Mean	SD	N^a	Mean	SD	N^a
1	0.30	1.26	162	0.42	1.28	175
2	0.17	0.90	15	0.83	0.98	20
3	0.20	1.10	30	0.45	1.12	32
4	0.17	1.38	27	0.42	1.36	25

[a]Number folllowed.
Source: Yudkin et al. (1991).

1" analysis of variance. The application of a random-effects model is "Model 2" analysis of variance. Mixed models are also possible. In this book, methods to estimate a summary estimate of effect in a two-group comparison based on a fixed-effects model (Model 1) will be described.

Table 8-1 gives the formulas to carry out an analysis with continuous measures. The first step in the analysis is estimating a summary measure of effect, the weighted grand mean. Next, a statistic, Q, referred to the chi-square distribution, is calculated and used to test the hypothesis of homogeneity of effect. If there is no statistical evidence of lack of homogeneity, a 95% confidence limit for the summary estimate of effect is calculated.

8.1.3 Application of the Method

EXAMPLE Table 8-2 presents data from a meta-analysis of the effect of azathioprine treatment on progression of disability in patients with multiple sclerosis (Yudkin et al. 1991). All of the studies eligible for inclusion in the meta-analysis used the Kurtzke Disability Status Scale to measure the effect of azathioprine. The following are the steps used to calculate a summary estimate of the effect of azathioprine on disability, a statistic to test the hypothesis that effects are homogenous, and an estimate of the 95% confidence interval for the summary estimate of effect. We begin by estimating the summary mean:

1. Estimate the mean difference between treatment and control for each study where

$$mean_i = mean_{ci} - mean_{ei}$$

Study 1: mean $_1$ = 0.42 − 0.30 = 0.12
Study 2: mean $_2$ = 0.83 − 0.17 = 0.66
Study 3: mean $_3$ = 0.45 − 0.20 = 0.25
Study 4: mean $_4$ = 0.42 − 0.17 = 0.25

2. Estimate the pooled variance for each study where

$$\text{variance}_{pi} = \left(\frac{SD_{ci}^2}{n_{ci}}\right) + \left(\frac{SD_{ei}^2}{n_{ei}}\right)$$

Study 1: $\text{variance}_{p1} = \dfrac{(1.28)^2}{175} + \dfrac{(1.26)^2}{162} = 0.019$

Study 2: $\text{variance}_{p2} = \dfrac{(0.98)^2}{20} + \dfrac{(0.90)^2}{15} = 0.102$

Study 3: $\text{variance}_{p3} = \dfrac{(1.12)^2}{32} + \dfrac{(1.10)^2}{30} = 0.080$

Study 4: $\text{variance}_{p4} = \dfrac{(1.36)^2}{25} + \dfrac{(1.38)^2}{27} = 0.145$

3. Calculate a weight for each study where

$$\text{weight}_i = \frac{1}{\text{variance}_{pi}}$$

Study 1: $\text{weight}_1 = \dfrac{1}{0.019} = 52.63$

Study 2: $\text{weight}_2 = \dfrac{1}{0.102} = 9.80$

Study 3: $\text{weight}_3 = \dfrac{1}{0.080} = 12.50$

Study 4: $\text{weight}_4 = \dfrac{1}{0.145} = 6.90$

4. Calculate the product of the weights and the mean difference for each study:

$$\text{product}_i = \text{weight}_i \times \text{mean}_i$$

Study 1: $\text{product}_1 = 52.63 \times 0.12 = 6.316$
Study 2: $\text{product}_2 = 9.80 \times 0.66 = 6.468$
Study 3: $\text{product}_3 = 12.50 \times 0.25 = 3.125$
Study 4: $\text{product}_4 = 6.90 \times 0.25 = 1.725$

5. Calculate the sum of the products:

$\text{sum of products}_i = 6.316 + 6.468 + 3.125 + 1.725 = 17.634$

6. Calculate the sum of the weights:

$\text{sum of weight}_i = 52.63 + 9.80 + 12.50 + 6.90 = 81.83$

7. Calculate the summary mean where

$$\text{mean}_s = \frac{\text{sum } (\text{weight}_i \times \text{mean}_i)}{\text{sum weight}_i}$$

$$\text{mean}_s = \frac{17.634}{81.83} = 0.22$$

We will next calculate Q, a statistic to test the hypothesis of homogeneity of effects:

1. Calculate the square of differences between the mean differences for each study and the summary mean:

   Study 1: $(\text{mean}_1 - \text{mean}_s)^2 = (0.12 - 0.22)^2 = 0.010$
   Study 2: $(\text{mean}_2 - \text{mean}_s)^2 = (0.66 - 0.22)^2 = 0.194$
   Study 3: $(\text{mean}_3 - \text{mean}_s)^2 = (0.25 - 0.22)^2 = 0.001$
   Study 4: $(\text{mean}_4 - \text{mean}_s)^2 = (0.25 - 0.22)^2 = 0.001$

2. Calculate the product of the weight$_i$ and the differences for each study as calculated in step 1:

   Study 1: weight$_1$ $\times$ 0.010 = 52.68 $\times$ 0.010 = 0.527
   Study 2: weight$_2$ $\times$ 0.194 = 9.80 $\times$ 0.194 = 1.901
   Study 3: weight$_3$ $\times$ 0.001 = 12.50 $\times$ 0.001 = 0.013
   Study 4: weight$_4$ $\times$ 0.001 = 6.90 $\times$ 0.001 = 0.007

3. Calculate Q as the sum of the results of step 2:

   $Q = 0.527 + 1.901 + 0.013 + 0.007 = 2.448$

4. Q is distributed as chi-square with degrees of freedom equal to one less than the number of studies. Based on Q with 3 degrees of freedom, the null hypothesis that the studies are homogeneous is not rejected because $p > 0.05$. There is no statistical evidence of lack of homogeneity and it is appropriate to use the summary weighted mean to estimate effect size.

Finally, we estimate the 95% confidence interval as follows:

1. Estimate a 95% confidence interval for the summary mean where

   $$CI = \text{mean}_s \pm 1.96 \times \sqrt{\text{variance}_s}$$

   and

   $$\text{variance}_s = \frac{1}{\text{sum weight}_i}$$

   $$\text{variance}_s = \frac{1}{81.83} = 0.012$$
   upper bound = $0.22 + (1.96 \times \sqrt{0.012}) = 0.22 + 0.22 = 0.44$
   lower bound = $0.22 - (1.96 \times \sqrt{0.012}) = 0.22 - 0.22 = 0.00$

8.1.4   When Effect Size Is Measured on Different Scales

When studies have used different scales to measure effect, the first step is to obtain an estimate of effect size for each study in a common metric. This is generally done as follows:

$$d_i = \frac{(\text{mean}_e - \text{mean}_c)}{SD_{pi}}$$

where d_i is the common metric that measures effect size in the ith study, $mean_e$ is the mean in the experimental (or exposed) group, $mean_c$ is the mean in the control (or unexposed) group, and SD_{pi} is the pooled estimate of the standard deviation of the effect measure for each study. When the study involves a before–after comparison, $mean_e$ and $mean_c$ are mean differences.

Hedges (1982) provides the following formula to estimate a summary effect size for studies that compare two groups taking into account the size of the two groups:

$$d_s = \frac{\text{sum } (w_i \times d_i)}{\text{sum } w_i}$$

where d_s is the summary estimate of the difference in the effect size measured in a common metric, w_i is the weight assigned to each study, and d_i is the effect size, estimated as described previously. This weighted estimator of the summary effect size was shown by Hedges to be asymptotically efficient when sample sizes in the two groups are both greater than 10 and the effect sizes are less than 1.5.

The weight of each study is

$$w_i = \frac{1}{\text{variance}_i}$$

where variance_i is the variance of d_i.

If sample sizes are about equal in the two groups and they are both greater than 10, the weight of each study can be estimated as follows (Hedges 1982; Rosenthal and Rubin 1982):

$$\text{weight}_i = \frac{2N_i}{(8 + d_i^2)}$$

where N_i is the total sample size in the ith study and d_i for each study is calculated as described previously.

A test of homogeneity can be carried out using a Q statistic that is wholly analogous to the Q described above, where

$$Q = \text{sum } [\text{weight}_i \times (d_i - d_s)^2]$$

Q is referred to a chi-square distribution with degrees of freedom equal to the number of studies minus 1.

Analogous to the situation described earlier, a 95% confidence interval for the summary estimate of effect size is estimated as

$$d_s \pm (1.96) \times \sqrt{\text{variance}_s}$$

where

$$\text{variance}_s = \frac{1}{\text{sum weight}_i}$$

Table 8-3 shows these formulas.

Table 8-3 Effects are measured as standardized differences: formulas to estimate the summary measure of effect, a statistic to test homogeneity, and 95% confidence interval

---

Summary measure of effect size

$$d_s = \frac{\text{sum (weight}_i \times d_i)}{(\text{sum weight}_i)}$$

$$d_i = \frac{\text{mean}_{\text{experimental}} - \text{mean}_{\text{control}_i}}{\text{SD}_{\text{pooled}_i}}$$

$$\text{weight}_i = \frac{1}{\text{variance}_i}$$

$$\text{variance}_i = \frac{8 + d_i^2}{2N_i}$$

where N_i is the total number of subjects in both groups

Test of homogeneity

$$Q = \text{sum [weight}_i \times (d_s - d_i)^2]$$

where the weight$_i$ are estimated as described above; Q is referred to the chi-square distribution with degrees of freedom equal to the number of studies minus 1

95% confidence interval

$$95\% \text{ C.I.} = d_s \pm (1.96 \times \sqrt{\text{variance}_s})$$

$$\text{variance}_s = \frac{1}{\text{sum weight}_i}$$

where the weight$_i$ are estimated as described above

8.1.5 Application of the Method

EXAMPLE Table 8-4 presents data from a meta-analysis of the effect of aminophylline in severe acute asthma that was done by Littenberg (1988). The 13 studies he identified as eligible for the meta-analysis were all studies of the effect of aminophylline on pulmonary function as measured by spirometry. However, the spirometry measures reported were not the same measures in each of the studies. The reported measures were converted to a common metric by dividing the mean difference in the experimental and control groups by an estimate of the pooled standard deviation for each study. The summary estimate of effect size and a 95% confidence interval are calculated as follows:

1. Calculate the weights for each study where

$$\text{weight}_i = \frac{2N_i}{8 + d_i^2}$$

$$\text{Study 1: weight}_1 = \frac{(2 \times 20)}{(8 + (-0.43)^2)} = 4.89$$

Table 8-4   Data from meta-analysis of the effect of aminophylline treatment in severe acute asthma

Reference	Total Number of Subjects	SD[a]	d[b]	Weight	Weight $\times$ d
Beswick et al. (1975)	20	0.76	−0.43	4.89	−2.10
Femi-Pearse et al. (1977)	50	320.00	−0.04	12.50	−0.50
Rossing et al. (1980)	48	0.65	−0.84	11.03	−9.27
Appel and Shim (1981)	24	0.42	−1.67	4.45	−7.43
Sharma et al. (1984)	29	0.22	−1.03	6.40	−6.59
Williams et al. (1975)	20	17.00	−2.41	2.90	−6.99
Tribe et al. (1976)	23	0.62	−0.08	5.75	−0.46
Evans et al. (1980)	13	110.00	0.26	3.22	0.84
Pierson et al. (1971)	23	2.10	2.93	2.77	8.12
Josephson et al. (1979)	51	6.30	0.51	12.35	6.30
Rossing et al. (1981)	61	0.50	0.72	14.32	10.31
Fanta et al. (1982)	66	0.67	0.03	16.50	0.50
Siegel et al. (1985)	40	0.58	−0.02	10.00	−0.20

[a]Standard deviation.

[b]Standardized difference = (improvement in treated group − improvement in control group)/SD.

Source: Littenberg (1988); table references cited there.

$$\text{Study 2: weight}_2 = \frac{(2 \times 50)}{(8 + (-0.04)^2)} = 12.50$$

and continue for all of the studies.

2. Calculate the product of the weight and the estimates of effect size for each study:

$$\text{product}_1 = (4.89) \times (-0.43) = -2.10$$
$$\text{product}_2 = (12.50) \times (0.04) = -0.50$$

and continue for all of the studies.

3. Calculate the sum of the weights:

sum of weights = 107.08

4. Calculate the sum of the products of the weight and the effect estimates:

sum of products = −7.47

5. Estimate the summary effect size where

$$d_s = \frac{\text{sum of products}}{\text{sum of weights}}$$
$$d_s = \frac{-7.47}{107.48} = -0.07$$

6. Estimate 95% confidence interval where

$$95\% \text{ C.I.} = d_s \pm (1.96 \times \sqrt{\text{variance}_s})$$

and

$$\text{variance}_s = \frac{1}{\text{sum weight}_i} = \frac{1}{107.48} = 0.009$$

upper bound $= -0.07 + (1.96 \times \sqrt{0.009}) = -0.07 + 0.186 = 0.116$
lower bound $= -0.07 - (1.96 \times \sqrt{0.009}) = -0.07 - 0.186 = -0.256$

The method is a method based on the assumptions of fixed effect. Calculations to test for heterogeneity of effect sizes are not shown.

8.1.6 Strengths and Limitations

The use of units of the standard deviation as a measure of the outcome of a comparative study is not accepted by all statisticians, even though the meta-analysis literature in the social sciences has focused on analysis of effect measures of this type. Greenland, Schlesselman, and Criqui (1987) give some examples where studies with identical results can spuriously appear to yield different results when the effect measures are converted to units of standard deviation.

In the social sciences, the effects of interventions are often measured using different instruments and the scales that result from this measurement process cannot be combined directly. It is impossible to do meta-analysis in these situations unless the effect measures are converted to a common metric. The example given in Section 8.1.4, which is from the medical literature, is similar. The effects of asthma treatment were measured using a variety of different measures of lung function derived from spirometry, and the only way to calculate a summary estimate of effect is to first convert the different measures of lung function to a common metric.

When there is no reason to convert the measures of effect to units of standard deviation, natural units should be used. This situation is the most common situation in the medical application of meta-analysis to describe the results of studies where effect size is measured on a continuous scale. Analyses based on units of the standard deviation are widely reported, and their results cannot be dismissed entirely.

8.2　ESTIMATING TREND

For many exposures, the presence or absence of a trend of increasing risk with increasing intensity of exposure is critical to the assessment of the causality of the association. Analyzing data from observational studies in terms only of "ever" and "never" exposed does not make full use of the information that is pertinent to the assessment of causality and it makes little sense considering biology. For most diseases, it is not reasonable to assume that extremely low-level, short-term exposures will affect risk. Including low-intensity exposures together with high-intensity exposures in a group called "ever use" obscures true associations. Even cigarette smoking, whose causal association with lung cancer is undisputed, does not show an association with risk at very low amounts for short periods of exposure. For example, smoking 1 cigarette per day for 30 weeks does not measurably increase the risk of lung cancer; if persons with this level of exposure are classified as smokers, a true association with lung cancer will be obscured.

The problems with defining "ever use" and "never use" dichotomies in the analysis of observational studies is especially acute in studies of common drug exposures, where there are many persons who have used a drug at least once. Aspirin, for example, is such a ubiquitous drug exposure that it should be obvious that studying "ever use" of aspirin in relation to almost any condition is a meaningless exercise. It is not much more meaningful to describe associations of disease with ever use of oral contraceptives, ever use of estrogen, or ever use of alcohol or caffeine.

Greenland (1987) describes how to use ordinary weighted least squares regression to estimate b, the slope of the trend of the odds of disease, with dose when information on the number of exposed cases and controls can be extracted from all of the reports in the meta-analysis. An extension of these methods to estimation of trend when all that can be extracted from the study reports are estimates of effect size (e.g., odds ratios and risk ratios) at various dose levels has also been described (Greenland and Longnecker 1992). Estimation of the betas and their variances requires the use of matrix algebra and a computer, and these calculations will not be described in detail. The interested reader is referred to Greenland and Longnecker.

Once the beta estimates of slope and their variances have been derived, estimating a summary slope and a 95% confidence interval for the summary estimate is straightforward. Table 8-5 gives the formulas, which should be familiar by now as they have the same general structure as other formulas in this and preceding chapters.

EXAMPLE: Table 8-6 shows the relative risk estimates for breast cancer in relation to amount of daily alcohol consumption as reported in the 16 studies that were eligible for the meta-analysis of alcohol and breast cancer done by Longnecker et al. (1988). The table also shows beta, which is the estimated increase in the log of the relative risk of breast cancer associated with an average daily alcohol consumption of 1 gram per day, along with the weights the studies carry in the estimation of the summary slope. These weights are calculated as the inverse of the variances. The betas and their variances were derived using the methods of Greenland and Longnecker (1992).

A summary estimate of the slope and a 95% confidence interval for the estimate were calculated with the formulas in Table 8-5 using steps that are exactly analogous to the steps given in Section 8.1. The summary estimate of beta is 0.00823. The value of Q, the statistic to test homogeneity, is 75.3. Q is referred to a chi-square distribution with 15 degrees of freedom. The

Table 8-5 Formulas to estimate the summary measure of slope, a statistic to test homogeneity, and 95% confidence interval for the summary measure

Summary measure of slope

$$\text{slope}_s = \frac{\text{sum (weight}_i \times \text{slope}_i)}{\text{sum weight}_i}$$

$$\text{weight}_i = \frac{1}{\text{variance}_i}$$

where the variance$_i$ are calculated as described by Greenland and Longnecker (1992) or as described by Greenland (1992) using ordinary least squares regression

Test of homogeneity

$$Q = \text{sum [weight}_i \times (\text{slope}_s - \text{slope}_i)^2]$$

where the weight$_i$ are estimated as described above; Q is referred to the chi-square distribution with degrees of freedom equal to the number of studies minus 1

95% confidence interval

$$\text{95\% C.I.} = \text{slope}_s \pm (1.96 \times \sqrt{\text{variance}_s})$$

$$\text{variance}_s = \frac{1}{\text{sum weight}_i}$$

where the weight$_i$ are estimated as described above

associated probability value is much less than 0.05, and the hypothesis of homogeneity is rejected. This means that the slopes are not the same, and the summary estimate of slope is suspect. Further exploration of the reasons for heterogeneity are in order.

8.3 MODELING IN META-ANALYSIS

The term "meta-regression" (Greenland 1987) has been used to describe analyses in which the characteristics of the studies are used to try to explain the magnitude of the effects observed in the individual studies or the deviation of the effect estimates in the individual studies from the summary estimate.

EXAMPLE: Phillips (1991) did a meta-analysis of studies of the sensitivity and specificity of tests for HIV seropositivity. There were 26 studies eligible for the meta-analysis. Information on the sensitivity and specificity of the HIV test was abstracted from each study, and each study was abstracted and classified according to year of publication (three categories), whether the sample test was a population with a low or high prevalence of HIV infection, and by study quality (high, medium, low). Table 8-7 presents the results of a regression analysis in which specificity of the HIV test as measured in each study was the dependent variable, and year of publication, HIV prevalence, and study quality were entered as predictors. The analysis

Table 8-6   Data from studies of association of grams of alcohol per day with breast cancer risk

Reference	Grams of Alcohol per Day	Estimated Relative Risk	Beta[a]	Weight[b]
Hiatt et al. (1984)	≤26	1.0	0.00434	164,000
	39–65	1.4		
	≥78	1.2		
Hiatt et al. (1988)	<13	1.2	0.0109	59,600
	13–26	1.5		
	39–65	1.5		
	≥78	3.3		
Willett et al. (1987)	<2	1.0	0.0284	31,400
	2–5	0.9		
	5–15	1.3		
	>15	1.6		
Schatzkin et al. (1987)	<1	1.4	0.118	441
	1–5	1.6		
	≥5	2.0		
Harvey et al. (1987)	<2	1.1	0.0121	54,200
	2–13	1.0		
	13–26	1.3		
	≥26	1.7		
Rosenberg et al. (1982)	<7	1.5	0.0870	1,860
	≥7	2.0		
Webster et al. (1983)	<7	0.9	0.00311	71,800
	7–21	0.9		
	21–28	1.1		
	29–36	1.1		
	36–43	1.0		
	≥43	1.1		

(*continued*)

showed that low prevalence of HIV was significantly associated with low test specificity.

Multivariate methods are a promising approach in meta-analysis, especially meta-analysis of nonexperimental studies. For some topics, the smallness of the number of studies may limit their usefulness in practice. The small number of studies available for the regression should not preclude the application of regression methods, but the number of explanatory variables should be kept small.

8.4   VOTE COUNTING AND RELATED METHODS

In the most simplistic method of "vote counting," the number of studies with statistically significant positive, negative, and null findings are tallied and the category with the plurality of votes is declared the "winner." There are no advocates

Table 8-6 (Continued)

Reference	Grams of Alcohol per Day	Estimated Relative Risk	Beta[a]	Weight[b]
Paganini-Hill and Funch (1982)	<13	1.0	0.0000	11,300
	≥26	1.0		
Byers and Funch (1982)	<1	1.1	0.00597	23,100
	1–3	1.0		
	4–11	1.1		
	>11	1.1		
Rohan and McMichael (1988)	<3	0.8	0.0479	2,378
	3–9	1.2		
	>9	1.6		
Talamini et al. (1984)	≤7	2.4	0.0389	16,900
	>7	16.7		
O'Connell et al. (1987)	≥2	1.5	0.203	112
Harris and Wynder (1988)	<5	1.0	−0.00673	56,900
	5–15	0.9		
	>15	0.9		
Le et al. (1984)	<11	1.0	0.0111	43,300
	11–22	1.4		
	23–34	1.5		
	>34	1.2		
LaVecchia et al. (1985)	≤39	1.3	0.0148	24,800
	>39	2.1		
Begg et al. (1983)	2–13	0.9	−0.000787	13,300
	>13	1.4		

[a]Increase in log relative risk for a gram/day of alcohol.

[b]1/variance.

Source: Longnecker et al. (1988) and Greenland and Longnecker (1992); table references cited in Longnecker et al. (1988).

Table 8-7 Results of regression analysis for 26 studies: specificity of HIV test in each study is the dependent variable[a]

Variable	Regression Coefficient	T	p
Year of publication	−0.023	−0.90	>0.05
Low-prevalence HIV[b]	0.114	2.54	<0.05
Study quality[c]			
High	−0.014	−0.20	<0.05
Low	−0.087	−1.38	<0.05

[a]After semilogarithmic transformation.

[b]High and mixed prevalence is referent.

[c]Intermediate quality is referent.

Source: Phillips (1991).

of this method, which is naive, has no statistical rationale, and can lead to erroneous conclusions (Hedges and Olkin 1980; Greenland 1987).

A more sophisticated version of vote counting is analysis based on the sign test. The sign test is based on the following reasoning. If the null hypothesis is true (i.e., there is no association between treatment and outcome or between exposure and disease), then one would expect that half of the studies would show a positive association and half would show a negative association.

EXAMPLE: In the meta-analysis of lung cancer and exposure to environmental tobacco smoke by the Environmental Protection Agency that was described in Chapter 1, 19 case-control studies were identified as eligible for the meta-analysis. Table 8-8 lists again the 19 studies along with the estimated relative risk of lung cancer in women exposed to environmental tobacco smoke. To do the sign test, each study is assigned a plus if the study estimated that the risk of lung cancer in exposed women was increased (i.e., estimated relative risk > 1.0) and a minus if the study estimated that the risk of lung cancer in exposed women was decreased (i.e., estimated relative risk < 1.0). Of the 19 studies, 16 have a plus sign and 3 have a minus sign. The null hypothesis is

Table 8-8 Data from meta-analysis of lung cancer in nonsmoking women exposed to environmental tobacco smoke

Reference	Estimated Relative Risk of Lung Cancer[a]	Sign	S[b]
Akiba, Kato, Blot (1986)	1.52	+	1.48
Browson et al. (1987)	1.52	+	0.61
Buffler et al. (1984)	0.81	−	−0.49
Chan et al. (1979)	0.75	−	−1.02
Correa et al. (1983)	2.07	+	1.52
Gao et al. (1987)	1.19	+	0.91
Garfinkel, Auerbach, Joubert (1985)	1.31	+	1.29
Geng, Liang, Zhang (1988)	2.16	+	2.19
Humble, Samet, Pathak (1987)	2.34	+	1.57
Inoue, Hirayama (1988)	2.55	+	1.50
Kabat, Wynder (1984)	0.79	−	−0.41
Koo et al. (1987)	1.55	+	1.56
Lam et al. (1987)	1.65	+	2.77
Lam (1985)	2.01	+	2.24
Lee, Chamberlain, Alderson (1986)	1.03	+	0.05
Pershagen, Hrubec, Svensson (1987)	1.28	+	0.91
Svensson, Pershagen, Klominek (1988)	1.26	+	0.57
Trichopoulos, Kalandidi, Sparros (1983)	2.13	+	2.55
Wu et al. (1985)	1.41	+	0.70

[a]In women exposed to environmental tobacco smoke compared with nonsmokers.

[b]S is the square root of the Mantel-Haenzel chi-square statistic with a positive sign for relative risk about 1.0 and minus for relative risk less than 1.0.

Source: Environmental Protection Agency (1990); table references cited there.

H_0: number of studies $+$ = number of studies $-$

Under the null hypothesis, the number of positive studies and the number of negative studies follow a binomial distribution with the probability of occurrence of a positive equal to $1 - \frac{1}{2}$, or $\frac{1}{2}$. The mechanics of the sign test are described in a number of textbooks (e.g., Armitage and Berry 1987). Using the methods described in these texts, one can estimate the probability of observing a 16:3 split strictly due to chance to be 0.002.

The sign test is a valid method of analysis of data where observations are individuals. It is not a recommended way to draw conclusions in a meta-analysis. First, the method yields no estimate of effect size. Second, it does not directly assess homogeneity of effect. Most important, it weights equally studies of all sizes and effects of all magnitudes.

8.5 STATISTICAL APPROACHES TO PUBLICATION BIAS

8.5.1 Overview

Chapter 4 described a quasi-statistical graphical method for assessing whether publication bias exists. Other statistical approaches to publication bias include methods that attempt to estimate the number of studies that would have to exist in order to explain the observed result of a meta-analysis and methods that attempt to adjust for unpublished studies. These methods are not recommended. They are described here along with the reasons for recommending against them.

8.5.2 Estimating the Number of Unpublished Studies

Rosenthal (1979) coined the term "file drawer problem" as a description of the problem of publication bias. According to his widely quoted description, the problem of publication bias arises because there are many studies whose results have been put into a "file drawer" without publication. Rosenthal describes a method for estimating the minimum number of unpublished studies with a null result that must exist in order to make the probability based on the observed (published) studies and the unobserved (unpublished) studies nonsignificant.

> *EXAMPLE:* A meta-analysis of 94 experiments examining the effects of interpersonal self-fulfilling prophecies yielded a probability value of $p < 0.001$. To reduce the p value to a barely significant level ($p = 0.05$), Rosenthal (1979) estimated that there would have to exist 3263 other, unpublished studies with an average null effect. The largeness of this number makes a strong intuitive appeal to the contention that unpublished studies do not explain the results of the meta-analysis based only on published studies.

This method is often cited and occasionally used. There are serious problems with the approach. First, in medical applications, one is rarely interested in knowing simply whether the result of the meta-analysis is or is not statistically significant. Second, the method uses normal theory and the Z statistic, which are not often the result of statistical analysis of medical data. Most important, the method assumes that the mean effect size of the unobserved studies is zero. That is, it assumes as true what is in doubt—that the unpublished studies taken together are null.

Orwin (1983) remedies the first problem by suggesting a method similar to Rosenthal's that is based on measures of the standardized difference in effect between a treatment and control group. This method estimates the number of studies whose overall mean is some number, say 0, that would be needed to bring the estimated difference between treatment and control in the observed studies to 0. Again, the estimated number of unobserved studies is sensitive to the assumption about the mean of the unobserved studies (Iyengar and Greenhouse 1988).

The idea of estimating how many unpublished studies there must be to explain the observed results of the meta-analysis as a way of assessing publication bias maintains its intuitive appeal. The statistical theory that would allow this to be done is not well developed. The methods that purport to allow these calculations to be made are all ad hoc, and the approach is not recommended for this reason.

8.5.3   Statistical Adjustment for Publication Bias

Hedges (1984) suggested an approach to publication bias which yields an estimate of a continuous measure "adjusted" for the effect of unobserved studies. The method, like that of Rosenthal (1979), is based on normal theory, and it assumes that errors are all normally distributed. An additional critical assumption, which makes it difficult to apply to the medical literature, is that all statistically significant studies are published and all nonsignificant studies are not published. This method is considered to be an improvement over the method of Rosenthal (see Iyengar and Greenhouse 1988), but it is not widely accepted. It is not recommended.

When sample sizes in each study are similar, Hedges and Olkin (1985) suggest an approach that uses binomial theory to estimate the effect, "adjusted" for unobserved studies. This approach also assumes that all statistically significant results are published and all nonsignificant results are not. This assumption is not tenable for medical applications, and the approach is essentially useless for this reason.

8.5.4   Other Methods

Iyengar and Greenhouse (1988) suggest a maximum-likelihood approach for estimating publication bias. Bayarri (1988) and Rao (1988) suggest that the problem of publication bias might best be dealt with using Bayesian approaches. Development of these methods is incomplete, and they are not covered further in this book.

9

Complex Decision Problems

Chapter 2 presented a simple decision analysis. The decision analysis was simple because only two alternative interventions were compared, because the events between the intervention and outcome required estimation of only a few probabilities, and because the outcome measure was a simple dichotomous measure—life or death. Decision analysis for most medical problems is more complex because it often involves comparison of more than one treatment or intervention, because the outcome of interest is not always a simple dichotomous outcome, and because the chain from treatment to outcome involves many events, requiring the estimation of many probabilities. This chapter begins the description of decision analysis for more complex situations by showing how to build and analyze decision trees with more complex outcomes and more complex intervening events. Chapter 10 describes the approach to estimating probabilities in a decision analysis. Incorporation of measures of the value of various outcomes to patients and society—utility analysis—is considered in Chapter 11. Sensitivity analysis is covered in Chapter 13.

Section 9.1 describes decision analysis with more than one outcome. Section 9.2 describes decision analysis involving comparison of more than two alternative treatments or interventions. Section 9.3 describes decision analysis involving many intervening events between intervention and outcome. Section 9.4 shows how to estimate life expectancy, which is commonly used as a measure of outcome in decision analysis. Section 9.5 discusses the use of Markov models to represent complex, time-related processes.

9.1   MORE THAN ONE OUTCOME

The outcome of most medical treatments is not a simple dichotomous outcome. Many medical treatments have side effects that need to be taken into account in making decisions about their net benefit and whether or not to recommend them. In addition, the beneficial and adverse consequences of many medical treatments and interventions include outcomes other than life and death.

In the simplest case, there are several mutually exclusive outcomes of an intervention. In this case, the decision tree is modified by including multiple boxes at the terminal or outcome node. This situation is shown for a hypothetical case in Figure 9-1. The probabilities of each outcome are estimated by literature review and recorded on the decision tree and analyzed by the process of folding back and averaging. In this situation, the decision analysis yields separate estimates of the value of each outcome in comparison with each alternative intervention.

> *EXAMPLE:* Measles can cause blindness as well as death. Figure 9-2 shows the decision tree based on the example used in Chapter 1 as modified to include the occurrence of blindness as an outcome. Blindness and death are mutually exclusive outcomes, and it is proper to record them at the terminal node of the decision tree.
>
> Based on review of the literature, it is determined that the likelihood of

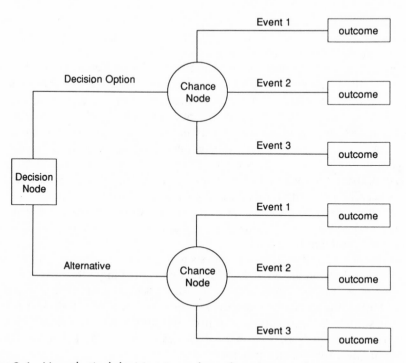

Figure 9-1   Hypothetical decision tree where there is more than one outcome.

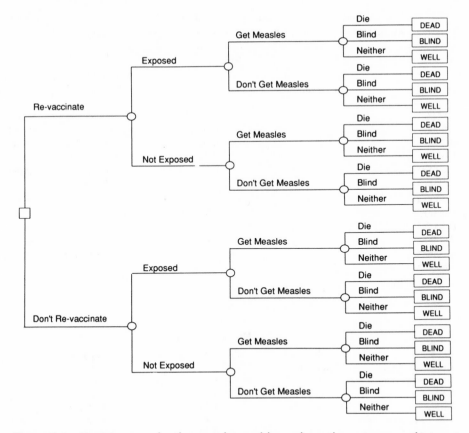

Figure 9-2 Decision tree for the measles problem where the outcomes of interest in the analysis are death, blindness, and remaining well.

blindness following measles is 45 cases per 100,000 cases of measles. The likelihood of blindness in the absence of measles is assumed to be 0.0.

Table 9-1 shows the numbers used to analyze the decision tree, again illustrating the analysis as a spreadsheet to simplify understanding of the calculations that are done to analyze the tree. In analyzing the decision tree, for each row, the product of the probability values in each column is computed. The expected number of deaths for the revaccination arm is estimated by summing the entries in the product column for the rows labeled with death in the upper part of the table. The expected number of cases of blindness in the revaccination arm is estimated by summing the entries in the product column for the rows labeled blindness in the upper part of the table. Thus, the expected number of deaths is just as it was in Chapter 2:

$$0.000023 + 0.000000 + 0.000000 + 0.000000 = 0.000023$$

The expected number of cases of blindness is

$$0.00045 + 0.000000 + 0.000000 + 0.000000 = 0.000045$$

Table 9-1　Calculations to show results of decision analysis for revaccination versus no revaccination where measles has three outcomes: death, blindness, well

| | Revaccination | | |
Product	Probability of Exposure	Probability of Getting Measles	Probability of Outcome	
0.000023	0.2	0.05	0.0023	die
0.000045	0.2	0.05	0.0045	blind
0.009932	0.2	0.05	0.9932	well
0.000000	0.2	0.95	0.0000	die
0.000000	0.2	0.95	0.0000	blind
0.190000	0.2	0.95	1.0000	well
0.000000	0.8	0	0.0023	die
0.000000	0.8	0	0.0045	blind
0.000000	0.8	0	0.9932	well
0.000000	0.8	1	0.0000	die
0.000000	0.8	1	0.0000	blind
0.800000	0.8	1	1.0000	well

Sum for deaths　　Sum for blind
0.000023　　　　　0.000045

| | No Revaccination | | |
Product	Probability of Exposure	Probability of Getting Measles	Probability of Outcome	
0.000152	0.2	0.33	0.0023	die
0.000297	0.2	0.33	0.0045	blind
0.065551	0.2	0.33	0.9932	well
0.000000	0.2	0.67	0.0000	die
0.000000	0.2	0.67	0.0000	blind
0.134000	0.2	0.67	1.0000	well
0.000000	0.8	0	0.0023	die
0.000000	0.8	0	0.0045	blind
0.000000	0.8	0	0.9932	well
0.000000	0.8	1	0.0000	die
0.000000	0.8	1	0.0000	blind
0.800000	0.8	1	1.0000	well

Sum for deaths　　Sum for blind
0.000152　　　　　0.000297

Differences between revaccination and no revaccination
Death 0.000129
Blind 0.000252

The expected number of cases of blindness and of death in the no-revaccination arm is estimated by summing the entries in the product columns for the rows corresponding to the relevant outcome entry. The expected number of cases of blindness is

$$0.000297 + 0.000000 + 0.000000 + 0.000000 = 0.000297$$

The expected number of deaths for the no-revaccination strategy is

$$0.000152 + 0.000000 + 0.000000 + 0.000000 = 0.000152$$

The strategies of revaccination and no-revaccination are compared for these two outcomes by subtracting the expected number of deaths and the expected number of cases of blindness. The expected number of deaths from measles is

$$0.000152 - 0.00023 = 0.000129$$

which is the same as in the prior example. The expected number of cases of blindness comparing revaccination with no-revaccination is

$$0.000297 - 0.000045 = 0.000252$$

Translated to numbers per 100,000 persons revaccinated, the revaccination strategy prevents 12.9 deaths from measles and 25.2 cases of blindness.

9.2 MORE THAN TWO ALTERNATIVE TREATMENTS OR INTERVENTIONS

More than two alternative treatments for a given condition may be available, or there may be more than one strategy for addressing a problem. In this case, the decision node of the decision tree has more than two arms. The expected outcome for each is calculated by the process of folding back and averaging, and the strategies are compared in relation to one another.

EXAMPLE: Revaccinating children is one strategy for addressing the problem of a measles epidemic. An alternative public health strategy to cope with the epidemic would be a strategy of excluding all children with any rash or fever from school for a two-week period. This "quarantine" strategy would be expected to decrease the likelihood of exposure to measles for children who remain at school and would prevent measles and its consequences for this reason.

Figure 9-3 is a decision tree depicting these three alternative courses of action—revaccination, quarantine, and no revaccination (do nothing). The decision tree has been simplified by removing the branches which have probabilities of 1.0 or 0, a process called "pruning."

Based on review of the literature, it is estimated that quarantine will decrease the likelihood of exposure to measles from 0.20 to 0.15. Table 9-2 shows the three alternatives with the relevant probabilities recorded. A new subtable has been added to represent the new decision alternative—quarantine—and the relevant probabilities are recorded in the table.

The estimates of the expected number of deaths from measles for the revaccination and no-revaccination (do-nothing) strategies do not change. The expected number of deaths from measles for the decision alternative, quarantine, is estimated by the process of folding back and averaging. The products of the probabilities in each row of the subtable are calculated.

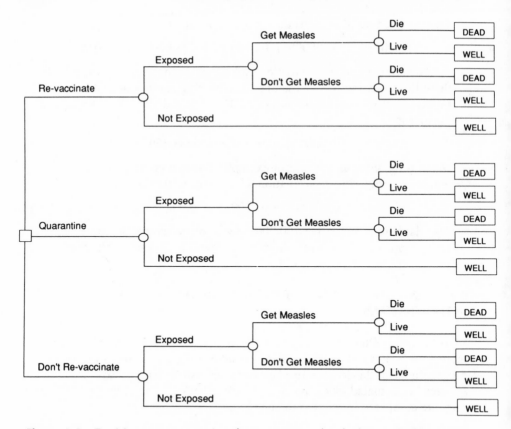

Figure 9-3 Decision tree comparing three strategies for dealing with the measles epidemic: revaccination, quarantine of infectious cases, and no revaccination.

Then, the expected number of deaths from measles is estimated by adding the entries in the columns corresponding to the rows labeled "die." The expected number of deaths is

$$0.000114 + 0.000000 + 0.000000 + 0.000000 = 0.000114$$

The comparison of the revaccination and no-revaccination strategies does not change. Revaccination is estimated to prevent 12.9 deaths per 100,000 children compared with no revaccination. The strategy of quarantine is compared with the strategy of no-revaccination (do nothing) by subtracting the expected numbers of deaths from measles as follows:

$$0.0000152 - 0.000114 = 0.000038$$

Interpreting these figures from the decision standpoint, the analysis shows that, while quarantine is expected to prevent 3.8 deaths per 100,000 compared with doing nothing, revaccination prevents 12.9 deaths per 100,000. Compared with doing nothing, the strategy of revaccination is superior to the strategy of quarantine.

Table 9-2 Calculations to show results of decision analysis comparing three options:revaccination, quarantine, and no revaccination

	Revaccination			
Product	Probability of Exposure	Probability of Getting Measles	Probability of Outcome	
0.000023	0.2	0.05	0.0023	die
0.009977	0.2	0.05	0.9977	don't die
0.000000	0.2	0.95	0.0000	die
0.190000	0.2	0.95	1.0000	don't die
0.000000	0.8	0	0.0023	die
0.000000	0.8	0	0.9977	don't die
0.000000	0.8	1	0.0000	die
0.800000	0.8	1	1.0000	don't die

Sum for deaths
0.000023

	No Revaccination			
Product	Probability of Exposure	Probability of Getting Measles	Probability of Outcome	
0.000152	0.2	0.33	0.0023	die
0.065848	0.2	0.33	0.9977	don't die
0.000000	0.2	0.67	0.0000	die
0.134000	0.2	0.67	1.0000	don't die
0.000000	0.8	0	0.0023	die
0.000000	0.8	0	0.9977	don't die
0.000000	0.8	1	0.0000	die
0.800000	0.8	1	1.0000	don't die

Sum for deaths
0.000152

	Quarantine			
Product	Probability of Exposure	Probability of Getting Measles	Probability of Outcome	
0.000114	0.15	0.33	0.0023	die
0.049386	0.15	0.33	0.9977	don't die
0.000000	0.15	0.67	0.0000	die
0.100500	0.15	0.67	1.0000	don't die
0.000000	0.85	0	0.0023	die
0.000000	0.85	0	0.9977	don't die
0.000000	0.85	1	0.0000	die
0.850000	0.85	1	1.0000	don't die

Sum for deaths
0.000114
Differences
Revaccination compared with no revaccination
0.000129
    12.9 deaths per 100,000
Quarantine compared with no revaccination
0.000038
    3.8 deaths per 100,000

9.3   MANY INTERVENING EVENTS

For most medical problems, the description of the pathway between a decision and its outcome involves many more intervening events than in the example that has been used in this book so far. The decision trees that result from the proper description of medical problems can be very complex. There are often many intervening events that themselves are determined by complex events.

> *EXAMPLE:* Chapter 1 briefly described a decision analysis that was done to try to inform clinical decisions about whether or not to give isoniazid prophylaxis routinely to HIV seropositive users of intravenous drugs. Figure 9-4 is the decision tree for this analysis. When the isoniazid arm of the tree is followed along its uppermost branches, the tree includes the occurrence or nonoccurrence of isoniazid toxicity. If isoniazid toxicity occurs, it is either fatal or nonfatal. If toxicity is nonfatal, then infection with tuberculosis either occurs or does not occur. If infection occurs, active tuberculosis may or may not result. If active tuberculosis occurs, it may be either fatal or nonfatal. The other branches of the tree can be similarly described.

Developing a decision tree that properly represents the problem that is posed is one of the most important challenges of decision analysis. The construction of complex decision trees is well-described in introductory textbooks by Weinstein and Fineberg (1980) and Sox, Blatt, and Higgins (1988), and these descriptions will not be repeated in this book. A software package that helps to construct complex decision trees, Decision Maker, has been developed and can be useful.

The methods for analyzing complex decision trees are a logical extension of the methods that were described in Chapter 2 and in Sections 9.2 and 9.3 and involve the process of folding back and averaging. The introductory textbooks by Weinstein and Fineberg (1980) and Sox et al. (1989) give examples of the analysis of complex decision trees, and such examples will not be provided in this book. For complex decision trees, the challenge of analysis is keeping track of the calculations. The specialized software, Decision Maker, is useful for this purpose. Otherwise, the help of a computer programmer may be necessary.

9.4   ESTIMATING LIFE EXPECTANCY

9.4.1   Overview

The outcome of interest in a decision analysis is often life expectancy and not just life or death. Since everyone must ultimately die, it is obvious that life expectancy is the most appropriate measure of the effect of an intervention whose most important effect is on survival.

Life expectancy is defined by actuaries as the average future lifetime of a person, and it is usually estimated for persons of a specific age, sex, and race. Actuarial methods to estimate life expectancy are based on specialized statistical life-table functions that rely on data on mortality rates specific for age, sex, and race.

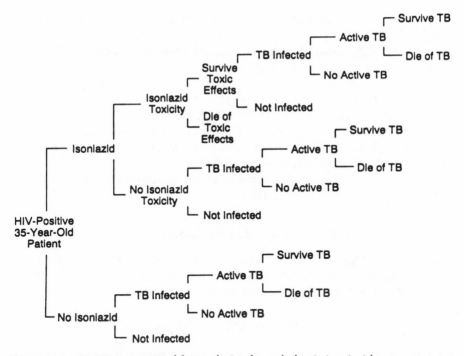

Figure 9-4 Decision tree used for analysis of prophylactic isoniazid versus no treatment in HIV-infected intravenous drug users. (Reproduced with permission from Jordan et al., *Journal of the American Medical Association,* 1991;265:2988.)

The age-, sex-, and race-specific mortality rates are based on death certificate data and census data.

Published tables of vital statistics describe the life expectancies of healthy persons. These published life expectancies are often all that is needed in a decision analysis, because the central task of the decision analysis is to estimate the effect of illness, with or without an intervention, on life expectancy.

In very rare cases, life expectancy in persons with an illness who have and have not undergone the intervention and its alternatives has been compared directly in a randomized trial or in a follow-up study. In these rare cases, the information on life expectancy can be used in a decision analysis without modification.

More often, available information on life expectancy in persons with a disease is in the form of overall mortality rates, five-year survival rates, or median survival. In general, the effect of interventions on life expectancy is measured as the relative risk or the odds of mortality in a given interval in those who have the intervention compared with those who do not. These kinds of information are not easily translated into information about life expectancy. For example, an intervention that halves the relative risk of death in a five-year follow-up interval does not double life expectancy. The effect on life expectancy of a disease that increases five-year survival by 20% is dependent on the age, sex, and race of the person, since life expectancy in the absence of the intervention is also dependent on these factors.

The estimation of life expectancy from information on overall mortality, five-year survival, median survival, and the relative risk of death in a given interval can be done with actuarial methods using information on age-, sex-, and rate-specific mortality. These actuarial methods require complex calculations that will not be described. A method for estimating life expectancy described by Beck, Kassirer, and Pauker (1982) and Beck et al. (1982) that requires only information on the age-, sex-, and race-specific life expectancy from a table of vital statistics and an estimate of the effect of the disease, treatment, or intervention on mortality has been widely used in decision analysis. The method, called the declining exponential approximation of life expectancy (DEALE), is simple to use, and it has been shown to closely approximate estimates of life expectancy based on actuarial methods (Beck, Kassirer, Pauker 1982).

9.4.2   Using the DEALE to Estimate Life Expectancy

Use of the DEALE assumes that survival follows a declining exponential curve. If this assumption is true, then life expectancy for a person of a given age, sex, and race can be estimated as the reciprocal of the mortality rate:

$$\text{life expectancy} = \frac{1}{\text{mortality}}$$

For a person of a specific age, sex, and race, this relationship can be used to estimate mortality from published life tables:

$$m_{asr} = \frac{1}{le_{asr}}$$

where m_{asr} is the average mortality rate of a person of a given age, sex, and race and le_{asr} is the life expectancy of a person of a given age, sex, and race as described in published life tables.

If an intervention decreases mortality by an amount m, then life expectancy for the person who has the intervention le_i is estimated as

$$le_i = \frac{1}{m_{asr} - m}$$

EXAMPLE: Imagine that an intervention decreases the probability of death by 0.001 per year. The problem is to determine the effect of the intervention on life expectancy in a 45-year-old woman.

1. Determine the average life expectancy at age 45 from a table of vital statistics:

$le_{asr} = 37.8$ years

2. Estimate the average mortality rate where

$$m_{asr} = \frac{1}{le_{asr}}$$

$$m_{asr} = \frac{1}{37.8} = 0.026 \text{ per year}$$

3. Estimate the mortality rate in those who have the intervention by subtracting the mortality rate caused by the intervention from the average mortality rate:

$$m_i = m_{asr} - m$$
$$m_i = 0.026 - 0.001 = 0.025 \text{ per year}$$

4. Estimate life expectancy in those who have the intervention where

$$le_i = \frac{1}{m_i}$$

$$le_i = \frac{1}{0.025} = 40.0 \text{ years}$$

5. Estimate the gain in life expectancy in those with the intervention compared with those without the intervention by subtraction:

$$\text{gain in life expectancy} = 40.0 - 37.8 = 2.2 \text{ years}$$

When the goal of the analysis is to estimate the effect of a disease on life expectancy, the same method can be used. In this case, excess mortality from the disease, m_e, is added to the mortality rate specific for age, sex, and race. That is, in step 3 described above:

$$m_d = m_{asr} + m \quad \text{and} \quad le_d = \frac{1}{m_d}$$

EXAMPLE: The effect of coronary heart disease on life expectancy in 45-year-old white women needs to be estimated. The excess mortality from coronary heart disease is 0.015 per year. The life expectancy of 45-year-old women with coronary heart disease is

$$m_d = m_{asr} + m = 0.026 + 0.015 = 0.041$$

$$le_d = \frac{1}{m_d} = \frac{1}{0.041} = 24.4 \text{ years}$$

Coronary heart disease reduces estimated life expectancy by 13.4 years (37.8 years − 24.4 years).

Excess mortality from various diseases and the effects of interventions on mortality per year are sometimes measured directly, and these equations are then directly applicable. More often, available information consists of overall mortality rate, five-year survival, or median survival in persons with the disease or having the intervention. Only a curve describing the survival of persons with the disease or having the intervention may be available. These measures of observed mortality are compound measures of mortality. All are composed of baseline mortality—the mortality expected in the general population plus either the excess mor-

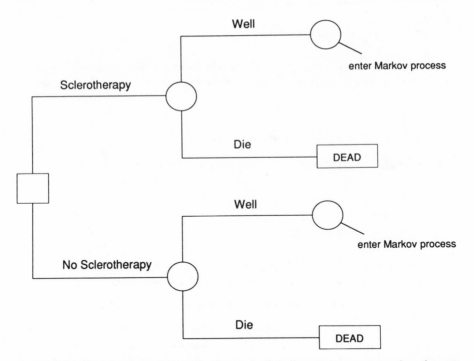

Figure 9-5 Decision tree for comparison of sclerotherapy versus no sclerotherapy for bleeding esophageal varices.

tality due to the disease or lower mortality due to the intervention. Before applying the DEALE, measures of compound mortality must be decomposed into baseline and excess mortality or baseline and saved mortality. Methods to decompose different kinds of compound measures of mortality so that they can be used to estimate life expectancy using the DEALE are described in detail by Beck et al. (1982), and they will not be described here.

9.5   MARKOV MODELS

9.5.1   Overall Goal

A Markov model is used in decision analysis to try to more accurately represent complex processes that involve transitions in and out of various states of health (Beck and Pauker 1983). Complex transitions are difficult to handle with decision trees analyzed using the methods that have been described. Markov models are used to attempt to capture the complexity of these transitions and incorporate this complexity into the decision analysis.

EXAMPLE: A decision analysis is being done to try to determine whether to recommend sclerotherapy for men with bleeding esophageal varices. Figure 9-5 is a decision tree drawn to represent this problem. The tree does not accurately represent the complexity of the problem of treating esophageal varices. Thus, a man with bleeding esophageal varices is initially ill and bleeding. He can either recover completely or die from the bleed. If he recovers, he is again at risk of bleeding from the varices. If another bleed from the varices occurs, he may recover or he may die from the bleed. Transitions in and out of states of health occur until the man dies of a bleed or from some other cause. The likelihood of death from other causes is high in men with bleeding esophageal varices. A decision analysis that tries to assess the effect of a treatment for bleeding esophageal varices on life expectancy should take into account the fact that transitions into and out of states of complete health occur and when they occur.

9.5.2 Application of the Method

9.5.2.1 Overview

There are four steps in a decision analysis that uses a Markov model to represent a process between the intervention and outcome. The first step is to determine the health states that will be modeled and to describe the ways in which transitions into and out of the health states will be allowed to occur in the model. The second step is to choose the length of the cycle that determines when transitions into and out of the various states that will be allowed. Third, the transition probabilities are estimated using the same methods that are used to estimate other probabilities in a decision analysis, as described in Chapter 10. Last, based on the estimates of the transition probabilities, the outcome with and without the intervention is determined by one of several methods.

9.5.2.2 Choose States and Transitions

It is common to represent the Markov process graphically. By convention, the states are defined as ovals or as circles. The time cycles are depicted on the left of the graph, and time runs downward on the graph. Arrows that link one state symbol with another state symbol are used to represent the allowed transitions between states in the model.

> *EXAMPLE:* Figure 9-6 is a graphic representation of the problem of bleeding esophageal varices that was described previously. In the figure, men are assumed to be bleeding from varices at time 0. In the interval from time 0 to time 1, all men with bleeding varices either become well or they can die. In the interval from time 1 to time 2, men who were well at time 1 can remain well, bleed again, or die from another cause. In the next interval, from time 2 to time 3, the men who bled again can recover (again) or die; the men who were well can remain well, rebleed, or die of another cause.

9.5.2.3 Choose Cycle Length

The cycle length is the amount of time that elapses between successive evaluations of outcome. It is chosen to reflect the underlying biological process that is being modeled, and it may be short (weeks) or long (years). Computationally, longer

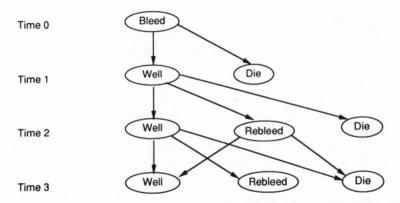

Time 0

Time 1

Time 2

Time 3

Figure 9-6 Graphical representation of Markov model of rebleeding after an initial episode of bleeding from esophageal varices.

cycles are less burdensome than shorter cycles, although the use of a computer program to carry out the Markov analysis makes consideration of the computational burden a relatively unimportant one.

> *EXAMPLE:* The cycle length chosen for the problem of bleeding varices is one year. That is, the outcome in the hypothetical cohort will be evaluated at the end of one-year cycles.

9.5.2.4 Determine Transition Probabilities

The next step is to determine the transition probabilities. The literature is usually the source of information on transition probabilities. Most available information about transitions between clinical states is expressed in the form of a rate and not in the form of a probability. A rate is the number of events per unit time; it can vary from zero to infinity. A probability is a quantity that is unitless (time is built into it); it takes values from zero to one. A rate r can be used to estimate a transition probability p of an event occurring over a time interval t based on the following formula:

$$p = 1 - e^{-rt}$$

where e is the base of the natural logarithm.

> *EXAMPLE:* In studies of men with bleeding esophageal varices, the rate of rebleeding is 51 per 100 per year. The transition probability from being well to rebleeding is
>
> $$P_{rebleed} = 1 - e^{-0.51} = 1 - 0.60 = 0.40$$
>
> This is the probability of a rebleed per year.
>   The probability of a rebleed per month is
>
> $$P_{rebleed} = 1 - e^{-0.51/12} = 1 - 0.96 = 0.04$$

Table 9-3 Rates per year and transition probabilities for sclerotherapy versus no sclerotherapy

Event	No Sclerotherapy		Sclerotherapy	
	Rate[a]	Transition Probability[b]	Rate[a]	Transition Probability[b]
Death at first bleed	0.91	0.60	0.69	0.50
Death of subsequent bleed	0.91	0.60	0.91	0.60
Rebleed	0.51	0.40	0.35	0.30
Death from other cause than bleed	0.51	0.40	0.51	0.40

[a]Per 100 per year.
[b]Per year.

Table 9-3 shows estimates of mortality at first bleed and subsequent bleeds, mortality from other causes, and the rate of rebleeding for men who do and do not undergo sclerotherapy as determined from a review of the literature. The transition probabilities per year calculated from these annual rates are also presented in the table.

9.5.2.5 Estimate Outcome

There are three main methods that can be used to provide information on life expectancy for the Markov process (Beck and Pauker 1983): Monte Carlo simulation, analysis of hypothetical cohorts of persons, and matrix algebra. The method for determining life expectancy for a Markov process that is conceptually the easiest is the method in which outcomes in hypothetical cohorts of individuals with and without the intervention are determined iteratively, usually until all members of the hypothetical cohort have "died." Monte Carlo simulation and the use of matrix algebra are somewhat more complex and will not be described here. The interested reader is referred to the description by Beck and Pauker (1983). The following example uses the hypothetical cohort method.

EXAMPLE: Table 9-3 gave the estimates of transition probabilities that are used in this example. Using these figures, in a hypothetical cohort of 100,000 men with bleeding esophageal varices at time T_0 who do not undergo sclerotherapy, it is estimated that 60,000 will die in the interval from T_0 to T_1 and 40,000 will be well. In the interval from T_1 to T_2, 16,000 of those who are well at the start of the interval will rebleed, 16,000 will die of other causes, and 8000 will remain well. In the next interval, 9600 of those who bled in the prior interval will die of the rebleed and 6400 will become well; of those well at the end of the last interval, 3200 will rebleed, 3200 will die of other causes, and 1600 will remain well. Calculations are repeated for this hypothetical cohort until all members of the cohort are estimated to have died, as shown in Table 9-4.

In a hypothetical cohort of 100,000 men who undergo sclerotherapy at the initial bleeding episode, sclerotherapy is assumed to decrease the prob-

Table 9-4   Markov process: calculations for hypothetical cohorts of men with bleeding esophageal varices who do or do not have sclerotherapy

	No Sclerotherapy			Sclerotherapy		
Time	Well	Bleed	Dead	Well	Bleed	Dead
0	0	100,000	0	0	100,000	0
1	40,000	0	60,000	50,000	0	50,000
2	8,000	16,000	16,000	15,000	15,000	20,000
3	8,000	3,200	12,800	10,500	4,500	15,000
4	2,880	3,200	5,120	4,950	3,150	6,900
5	1,856	1,152	3,072	2,745	1,485	3,870
6	833	742	1,433	1,417	824	1,989
7	464	333	778	755	425	1,061
8	225	186	386	396	227	557
9	119	90	202	210	119	294
10	59	48	102	111	63	155
11	30	24	53	59	33	82
12	16	12	26	30	18	44
13	9	6	13	16	9	23
14	4	3	8	9	4	11
15	0	0	7[a]	4	3	7
16	0	0	0	0	0	7[a]
Sum	62,495	124,996		86,202	125,860	
Average cycles[b]	0.62	1.25		0.86	1.26	

Life expectancy: no sclerotherapy $0.62 + 1.25 = 1.87$ years
Life expectancy: sclerotherapy $0.86 + 1.26 = 2.12$ years
   Difference $2.12 - 1.87 = 0.25$ year

[a]The tail has been truncated.
[b]Sum/100,000.

ability of death at the time of the first bleed from 0.60 to 0.50 and to decrease the probability of rebleeding in each subsequent interval from 0.4 to 0.3, but not to affect either the probability of death given a rebleed or the probability of death from other causes. In this hypothetical cohort, the numbers alive, bleeding, and well are as shown in Table 9-4.

Once the numbers of men in states of being well, rebleeding, and death for each cycle have been calculated, the total number of years spent in the states of well or rebleed is determined by summing the relevant columns. This sum is divided by the size of the hypothetical cohort to determine the average cycle length spent in each of the two living states. Life expectancy is estimated as the sum of the average cycles spent in living states. In the no-sclerotherapy cohort, the average number of years in the well state is 0.62. In the state of bleeding, it is 1.25 years. Both the well state and the bleeding state are living states, and life expectancy is estimated as the sum of the average cycles in these two states, or 1.87 years. In the sclerotherapy cohort, life expectancy is estimated to be 2.12 years, the sum of the average number of years spent in the well state (0.86) and the average number spent in the state of bleeding (1.26). The estimated gain in life expectancy from sclerotherapy is estimated as the difference in these two life expectancies, which is 2.12 − 1.87, or 0.25 year.

9.5.3 Markov Chains Versus Markov Processes

Markov models can assume either that the probabilities of transition are constant over time or that they vary. The first class of models are Markov chain models; the second class are Markov process models. Modeling a problem as a Markov process is required whenever the death occurs remote in time to the intervention, or the cohort "ages." The preceding example used a Markov chain model. In the example, it was not necessary to take age into account because the underlying mortality rate from causes other than bleeding in a cohort of men with bleeding esophageal varices is so high that the increasing mortality rate with age does not come into play in the estimate of life expectancy. Methods for incorporating transition probabilities that vary over time into a decision analysis based on a Markov model are provided by Beck and Pauker (1983).

9.5.4 Limitations

Use of a Markov model to represent a process assumes that the probability of a transition out of any given state is independent of the prior transitions. This assumption is a strong assumption, and violations can lead to erroneous conclusions. The assumption is a fairly tenuous one in many medical applications.

Additional limitations on the use of Markov models arise because of the unavailability of information that would allow accurate estimation of transition probabilities. Special studies to estimate these are rarely undertaken, and data to derive the estimates may not be easily obtainable.

10

Estimating Probabilities

Proper choice of estimates of the probabilities for a decision analysis is a critically important task. The uncertainties in the probability estimates need to be considered in drawing conclusions from the decision analysis. This chapter describes how to select probabilities and how to justify the choice of probabilities. It presents some of the methods for estimating uncertainty in the expected outcome.

Section 10.1 describes the overall goals of the process of selecting probabilities. Sections 10.2 and 10.3 describe the use of published sources to estimate probabilities. Section 10.4 discusses use of expert panels and Section 10.5 the use of subjective judgment and "guesses" for this purpose. Section 10.6 describes statistical and quasi-statistical methods that attempt to take the uncertainty in estimates of probability into account in the overall decision analysis. Sensitivity analysis is considered in Chapter 13.

10.1 OVERALL GOALS

The overall goal of the process of selecting the probabilities for a decision analysis is selection of probabilities that represent the truth. There are three important considerations in the gathering of information to fill in the decision tree. First, identification of information should be systematic. Second, the sources of information should be documented and, where there are alternative sources of information, a rationale for the choice should be provided. Third, effort should be expended to identify the best available information for each probability and for

Table 10-1 Probabilities used in a decision analysis of the effectiveness of prenatal screening and immunization for hepatitis B virus

Outcome	Baseline Probability	Range of Probabilities	Sources
Mother with +HBsAg	.002	.001–.15	Hoofnagle and Alter (1984); Advisory Committee on Immunization Practices (1985); Malison et al. (1985); Malecki et al. (1986)
HBsAG test			
Specificity	.980	0.96–.99	Holland (1985)
Sensitivity	.975	.95–.99	Holland (1985)
Probability of transmission	.425	.125–.90	Wong et al. (1984); Beasley et al. (1982)
Efficacy of immunization	.90	.85–.95	Wong et al. (1984); Beasley et al. (1982); Stevens et al. (1985)
Probability of acute hepatitis in neonates	.025	.02–.03	Delaplane et al. (1983); Sinatra et al. (1982)
Probability of death	.25	.25–.30	Beasley (1982); Beasley et al. (1981)

Source: Arevallo and Washington (1988); table references cited there.

the measure of outcome or the utility, or accurate information on the probability should be generated *de novo.*

When published studies are used as sources of probabilities, the validity of the studies that are the source of the probability estimates is a critical consideration. In addition, because complete agreement about the value of a probability is rare and because even probabilities based on large and definitive studies are associated with sampling error, the range of reasonable estimates of the probabilities needs to determined at the same time that the baseline estimate is determined. The upper and lower values of the probability are measures of the uncertainty in the baseline estimate. These values are used in sensitivity analysis and in statistical and quasi-statistical analyses that attempt to estimate the uncertainty in the measure of expected outcome. Justification for the choice of values considered to represent the range of reasonable estimates of the probability needs to be rigorous.

The process of selecting probabilities should generate a table that shows the best estimate of the probability, the upper and lower values of the probability that are considered reasonable, and the source of the baseline probabilities and the probabilities considered reasonable. The best estimates are called "baseline" estimates. An analysis that uses the best estimates is called the baseline analysis.

EXAMPLE: Table 10-1 is a table of the probabilities used in a decision analysis of the effectiveness of prenatal screening and immunization for hepatitis B virus done by Arevalo and Washington (1988). This table is an excellent example of the final product of the process of identifying probabilities for a decision analysis.

Probabilities used in decision analysis and cost-effectiveness analysis are usually derived from the published literature. Sometimes there is only one source of information on a particular probability, and this fact is known. Sometimes a source is used to estimate a probability for specific reasons: the source is authoritative, or it is the largest or the best study. In other situations, the literature search procedures described in Chapter 4 can be used to identify systematically information on the probability. If the literature search identifies more than one source of information on the probability, either meta-analysis or ad hoc procedures for aggregating the data can be used to yield probability estimates.

In some cases there is no published source of information on a probability. In these situations the analysis may rely on experts to estimate probabilities. It is possible to use educated "guesses" or to rely on subjective judgments. Last, information to estimate a probability can be generated by special studies done specifically for this purpose.

10.2 RELYING ON SELECTED PUBLISHED SOURCES OF INFORMATION ON PROBABILITIES

10.2.1 Using a Source Because It Is the Only Published Source

Sometimes only one source of information to estimate a probability is available. In these cases, it seems obvious that this source must be used in the analysis.

EXAMPLES: Schulman et al. (1991) did an analysis of the cost effectiveness of low-dose zidovudine (azidothymidine, AZT) therapy for asymptomatic patients with human immunodeficiency virus (HIV) infection. To estimate the effect of low-dose zidovudine on the clinical course of HIV infection, they used data from the AIDS Clinical Trials Group Protocol 019, which was the only study of the topic that had been completed at the time of their analysis.

Ransohoff et al. (1983) did a decision analysis comparing prophylactic cholecystectomy with expectant management for silent gallstone disease. At the time of their analysis, only one study had ascertained systematically gallstones and then followed patients with silent gallstones for a long period of time.

The availability of only one source of information is usually known to the contemporary readers of the results of the analysis. Other studies may be done after the decision analysis is published, and later readers may not have the time frame of the analysis firmly in mind. For this reason, it is a good idea to mention that the publication has been selected to estimate the probabilities because it is a sole source.

When there is only one published source of information about a probability, the range for the probability can be based on the 95% confidence interval from that study.

10.2.2   Using a Source Because It Is Authoritative

Sometimes an information source has been widely used by others and is considered to be authoritative.

>   *EXAMPLE:*   Goldman et al. (1991) did a cost-effectiveness analysis of treatment with HMG-CoA reductase inhibitors (e.g., lovastatin) in the primary and secondary prevention of coronary heart disease. One goal of the analysis was estimation of the effectiveness of treatment for men and women in several age groups. To make these calculations, estimates of the incidence of coronary heart disease in men and women were required. Data on the relative risk of coronary heart disease in men and women by age derived from the Framingham Heart Study were used because this source is considered authoritative.

When a probability estimate is chosen because the source is considered authoritative, this explanation for the choice of the source should be described in the study report.

The range should include other studies of the same topic that were not considered authoritative. Alternatively, if a 95% confidence interval for the probability is presented in the authoritative source or can be calculated based on available data, it can be used.

10.2.3   Using a Source Based on Its Size or Quality

Sometimes one study is so much larger than any other study that it can be seen a priori that the small studies contribute no useful information. There are instances when one study is so much better than any of the others that it is justified to use the information only from it.

>   *EXAMPLE:*   Heckerling and Verp (1991) did a decision analysis comparing amniocentesis and chorionic villus sampling for prenatal genetic testing. They used published information from one very large study to estimate the prevalence of a chromosomal abnormality at birth in an infant born to a 35-year-old woman.

When quality or size is the reason for choosing a study, this explanation belongs in the report.

The upper and lower estimates for a probability that is based on a single large study can be based on the 95% confidence interval from the study.

10.2.4   Using a Source Because It Is Representative

When results of an analysis will be generalized to a particular population, it is a good idea to use information that is representative of that population in the analysis. When an analysis focuses on a particular intervention and there are studies specifically of that intervention, the analysis should rely on data from studies of that intervention whenever possible.

> *EXAMPLES:*   Oster and Epstein (1987) did a cost-effectiveness analysis of cholestyramine for the prevention of coronary heart disease. They used information from the Lipid Research Clinics Coronary Primary Prevention Trial (1984A, 1984B) as the source of their estimate of the amount of reduction in the risk of coronary heart disease associated with use of the drug, because this study was a study of cholestyramine. This contrasts with the cost effectiveness of HMG-CoA reductase inhibitors for prevention of coronary heart disease done by Goldman et al. (1991). This group used pooled data from six studies of the effect cholesterol-lowering drugs to estimate the amount of reduction in the risk of coronary heart disease that would be expected with use of HMG-CoA reductase inhibitors, because no studies estimating the reduction in the risk of coronary heart disease in users HMG-CoA reductase inhibitors had been done at that time.
>
> In the analysis by Schulman et al. (1991) of low-dose zidovudine therapy for asymptomatic patients with human immunodeficiency virus discussed earlier, information from the AIDS Clinical Trials Group Protocol 019 was used not only to estimate the effects of zidovudine on outcome in patients with asymptomatic HIV infection, but also to estimate incremental number of physician visits and tests needed by those treated. The authors reasoned that use of physician services and tests in persons in the study would be representative of use in clinical practice.

10.3   AGGREGATING INFORMATION FROM MULTIPLE PUBLISHED SOURCES

When there is more than one source of information on a probability, a meta-analysis done by others can be used as the source of the probability estimate, or a formal meta-analysis can be carried out as part of the study. Information can be aggregated in an ad hoc fashion that falls short of formal statistical meta-analysis.

10.3.1   Meta-Analysis

When multiple studies of a given topic exist, meta-analysis is the preferred method for using the information to derive a probability for several reasons. First, the systematic nature of the information retrieval process for meta-analysis maximizes the likelihood of obtaining an unbiased estimate of the probability. Second, meta-analysis formally takes into account the size of the studies and insulates the decision analysis from criticisms of arbitrariness. Last, use of meta-analysis enhances reproducibility.

> *EXAMPLE:*   Oster, Tuden, and Colditz (1987) did an analysis comparing the cost effectiveness of different methods of prophylaxis against deep vein thrombosis for patients undergoing major orthopedic surgery com-

pared with observation. To estimate the percentage of patients who would get deep vein thrombosis without any treatment, they did a meta-analyses of 16 randomized trials of prophylaxis that had used a placebo or no-prophylaxis control group. Rates of deep vein thrombosis in the various treatment groups were similarly estimated by meta-analysis of randomized trials that involved the treatment as one of the arms in the trial.

10.3.2   Ad hoc Aggregation

The data from a group of studies assembled systematically or informally can be aggregated by averaging or by some "pooling" procedure that falls short of formal statistical meta-analysis. Averaging and other nonstatistical methods of pooling are very common in decision analysis and cost-effectiveness analysis.

> *EXAMPLE:*   In Arevalo and Washington's analysis of the cost effectiveness of prenatal screening and immunization for hepatitis B virus (1988), the efficacy of the hepatitis B vaccine was one of the probabilities in the decision analysis. The rate used in the baseline analysis was taken to be the mean of the rates given in three publications on the topic.

Averaging does not take into account the size of the studies, and all studies are given equal weight in calculating the average. The quality of the studies is also not taken into account. Although it is common to average measures from different studies to obtain a probability measure for a decision analysis, this method is not recommended. When there is more than one source of information on a probability, the method of pooling should take the variance of the measures from the individual studies into account.

10.4   EXPERT PANELS AS SOURCES OF PROBABILITY ESTIMATES

Experts and expert panels are often mentioned as a source of data on probabilities in clinical decision analysis (Weinstein and Fineberg 1980; Sox et al. 1988). Examples of the use of expert panels to derive probability estimates for a decision analysis aimed at formulation of clinical or public policy could not be identified. Carefully constituted expert panels may be acceptable, but this method of gathering information for a decision analysis should probably be used only when other sources do not exist. If an expert panel is used as the source of information on a probability used in a decision analysis, the choice of experts needs to be justified. The methods for soliciting the information from the experts need to be described.

10.5   PERSONAL EXPERIENCE AND "GUESSING" TO ESTIMATE PROBABILITIES

Personal experience and "guessing" have been described as methods for deriving probabilities in *clinical* decision analysis—decision analysis done at the bedside to help guide the management of an individual patient (Weinstein and Fineberg 1980; Sox et al. 1988). Published examples of the use of subjective experience and guesses in decision analysis aimed at formulation of clinical and public policy could not be identified. Probabilities based on guesses and subjective experience are likely to be criticized in the peer review process, and studies that use such methods probably will not be highly regarded by the scientific community. The use of personal experience and guessing are not recommended approaches to estimating probabilities outside the clinical setting.

10.6   ACCOUNTING FOR UNCERTAINTY IN PROBABILITY ESTIMATES

Most decision analyses involve estimation of many different probabilities. Errors in these uncertainties propagate when the probabilities are multiplied. When the decision model is complex, even small errors in the individual probability estimates can lead to large errors in the final result of the decision analysis.

Sensitivity analysis is the method that is used most often to try to evaluate the effect of uncertainty in the probability estimates on the conclusions of the decision analysis. More generally, sensitivity analysis attempts to evaluate the effect of various assumptions made in an analysis on the conclusions that are reached. Sensitivity analysis is an essential component of both decision analysis and cost-effectiveness analysis. The topic is covered in a separate chapter, Chapter 13, because it is the most commonly used method for assessing uncertainty in decision analysis and cost-effectiveness analysis.

Unfortunately, sensitivity analysis in which three or more probabilities are varied at the same time is difficult to do, and it is difficult to make the results comprehensible. Because of this limitation of sensitivity analysis, statistical and quasi-statistical methods to derive a single summary estimate of the uncertainty in the outcome measure taking into account the uncertainties in all of the probability estimates used in the decision analysis have been developed (Rifkin 1983; Doubilet et al. 1985; Critchfield and Willard 1986). Each of these methods tries to estimate some kind of "confidence range," which is meant to be analogous to the confidence interval that is derived from analysis of data subjected to classical statistical analysis. This estimate is an estimate of the uncertainty in the difference in outcome between the two alternatives examined in the decision analysis. These methods are described and critiqued.

10.6.1   Description of the Methods

For each of the methods, the probability and outcome measures are assumed to be variables with an associated probability density function. The probability density function for each variable has a mean and a standard deviation. The mean of the probability density function for each variable is equal to the probability estimate. The standard deviation is assumed to reflect the degree of uncertainty in the variable. For simple decision trees, the expected outcome of each decision option and a probability density function for the expected outcome are calculated directly using this information.

For complex decision trees, precise calculations of the probability density function for the outcome measure are tedious and often impossible. Thus, in practice, the probability density function of the expected outcome is estimated using Monte Carlo simulation. The decision tree is repeatedly analyzed by picking values of the probabilities and the outcome at random based on the assumed underlying probability density function. At each iteration, the result of the analysis is recorded. The mean of the results of the many iterations estimates the expected outcome. The results of the many iterations are assumed to be normally distributed, and standard normal theory is used to estimate a 95% "confidence range" for the expected outcome.

Specifying probability density functions for all of the probabilities and the outcome measure is a critical part of application of these methods. Doing this can be very time-consuming. For many variables, there are no empiric data on the probability density function, and the choice of the probability density function is a guess. Alternatively, the probability is assumed to have a distribution that is convenient. Doubilet et al. (1985), for example, describe a method for estimating the probability density functions for each of the probabilities in a decision analysis that simplifies application of the Monte Carlo method. They suggest assuming that the probability density function for each probability in the decision analysis follows a logistic-normal distribution. Details of the method are presented in their publication.

10.6.2 Bayesian Methods

Eddy and colleagues (Eddy 1989; Eddy, Hasselblad, Shachter 1990A, 1990B) described a method for synthesizing evidence that is based on Bayesian statistics. The method is called the "confidence profile method." Eddy and colleagues frame the method as a method for estimating uncertainty in a meta-analysis. It is more clearly related to decision analysis and to the broader problem of estimating uncertainty when multiple sources of information are used to estimate the outcome for a clinical problem. Like the methods described previously, the confidence profile method yields a quantity conceptually analogous to the 95% confidence interval based on classical statistics. The methods theoretically can be used to take into account the bias in individual studies as well as the uncertainty in the estimates of individual study parameters.

The full use of these methods requires a fair amount of subjectivity in specifying a prior distribution to represent all of the basic parameters in the model (or to decide to use a noninformative prior distribution). The problem of identifying and measuring bias in individual studies is not addressed by the methods. Adjusting for bias using the methods may require more information from individual studies than is available to the analyst.

10.6.3 Critique of the Methods

None of these statistical and quasi-statistical methods for estimating the uncertainty in the estimate of the expected outcome has been widely applied. The estimation of a probability density function for each probability in the analysis is a formidable undertaking even when an expert panel or a single judge is used for this purpose. The validity and reliability of estimates of the probability density function provided by an expert panel or a single judge are uncertain. The final estimate of the uncertainty of the expected outcome is dependent on the assumptions about the probability density functions of the individual probabilities in the decision analysis.

The suggestion by Doubilet et al. (1985) that probability density functions for all of the probabilities included in a decision analysis be assumed to be logistic-normal is based on the convenience and mathematical tractability of the logistic-normal distribution and not on empiric or theoretical data to support its use. The assumption that expected outcome of the Monte Carlo simulation estimating expected outcome follows a normal distribution and that standard normal theory applies is unproven.

The Bayesian approach of Eddy and colleagues (Eddy 1989; Eddy, Hasselblad, Shachter 1990A, 1990B) appears promising for its applications in both meta-analysis and decision analysis. However, the descriptions of the method are difficult to follow. Empirical evaluation is lacking.

Sensitivity analysis is a well-accepted, straightforward method for attempting to assess the amount of uncertainty in a decision analysis. Although sensitivity analysis does not yield a quantitative estimate of the amount of uncertainty in the estimate of expected outcome, the methods that claim to yield such estimates are poorly developed. In the absence of better empirical and theoretical work to support their use, these methods should be used cautiously.

11

Utility Analysis

In the examples of decision analysis presented so far, the outcome of interest in the decision analysis has been life or death, or some measure closely related to life or death, such as years of life lost, years of life saved, or life expectancy. The ability of an intervention to prevent death or to prolong life narrowly frames the goal of many medical treatments and interventions. Many treatments decrease the likelihood of disability but not death, and many treatments increase life expectancy but at the cost of impaired function. Attempts to incorporate into decision analysis effects of interventions on outcomes other than life or death, and the value or preferences of individuals or society for these outcomes, is called "utility analysis." This chapter describes the measurement of utilities and how to use these measures to do utility analysis.

Section 11.1 discusses the concept of utility in more detail. Section 11.2 describes the important conceptual issues in the measurement of preferences for health states. Section 11.3 describes how to measure preferences for health states in practice and some of the limitations of the measures. Section 11.4 describes how to incorporate measures of preference into a decision analysis to derive quality-adjusted life expectancy as the outcome of a decision analysis. Section 11.5 discusses some problems and concerns about measurement of health preferences.

11.1 THE CONCEPT OF UTILITY

In decision analysis, "utility" refers to the desirability or preference that individuals or societies have for a given outcome (Torrance 1987). Preferences are the levels of satisfaction, distress, or desirability that people associate with the health outcome (Froberg and Kane 1989A). Decision analysis increasingly concerns itself with the measurement of preferences and with the incorporation of these measures into the analysis. A decision analysis that incorporates measures of the preferences of individuals or society is called "utility analysis."

When a decision analysis focuses on the difference in the probability of living or on life expectancy as the outcome of the intervention and its alternatives, life and life expectancy are implicitly considered to be valued, and people are assumed to prefer a greater probability of living and enhanced life expectancy. The need for incorporating measures other than life and life expectancy into a decision analysis follows from the observation that many medical interventions do not affect life or life expectancy and some treatments alter the chances of living or dying but are themselves associated with morbidity. For example, treatments for chronic arthritis, back pain, and headache affect pain and disability but not life expectancy; chemotherapy for acute lymphocytic leukemia in children causes temporarily disabling and unpleasant side effects but increases life expectancy.

In many cases, decision analysis and utility analysis are used to estimate the effect of the intervention on quality adjusted life expectancy. An outcome may be preferred not because of its effect on a measure of health, but for other reasons. The broad concept of utility analysis accommodates the need to assess the effects of interventions and their alternatives on outcomes that are preferred or valued for any reason, even reasons other than their effects on health or health status. For example, vaginal delivery of women with normal pregnancies may be preferred on a priori grounds. A utility analysis could be done to assess the effect of an intervention on the likelihood of vaginal delivery. In this case, the likelihood of vaginal delivery would be the utility measure. Because the measurement of preferences for health states and the incorporation of these measures into decision analysis in the form of quality-adjusted life expectancy has such a prominent place in the literature of decision analysis and cost-effectiveness analysis in medicine, this chapter focuses on utilities that reflect preferences for health states.

11.2 CONCEPTUAL ISSUES IN THE MEASUREMENT OF PREFERENCES FOR HEALTH STATES

11.2.1 Overview

The overall goal of measuring preferences for health states is to construct a reliable and valid numerical representation of these preferences. To be used in decision analysis, this representation is put in the form of a numerical scale that ranges from 0 to 1.0, where 0 represents the least preferred health state and 1.0 the most preferred state. The scale values assigned to the health states are the values used to "adjust" life expectancy—yielding a measure of quality-adjusted life expectancy. The scale values can also be assigned directly to the outcomes of the interventions, or they can themselves serve as the outcomes.

11.2.2 Validity and Reliability

A technique to measure preferences for health states should be reliable and valid. A measure is reliable if the same phenomenon can be measured a second time with the same result. A measure is valid if it measures the phenomenon it claims to measure.

For measures of preferences for health states, evaluation of three types of reliability is pertinent—internal (intrarater) reliability, test–retest reliability, and interrater reliability. Internal (intrarater) reliability refers to the similarity of replicate measures on the same subject at the same time. Test–retest reliability refers to the similarity of the same measurements made at different times on the same person. Interrater reliability refers to the similarity of the measure made by two or more raters.

Evaluation of two types of validity is pertinent to the construction of measures of preferences for health states: criterion validity and construct validity (Spitzer 1987). Evaluation of criterion validity involves the comparison of the measure with a well-accepted measure of the concept being measured. To evaluate criterion validity, there must be a "gold standard." Evaluation of construct validity requires relating the measure to a concept or a group of concepts and testing the ability of the measurement to predict events or behaviors that grow out of an understanding of these concepts (Torrance 1976). A measure of preferences for health states would have construct validity if it predicted the actual choices made by individuals or groups of individuals. In practice, construct validity is rarely evaluated in the health preferences literature. Rather, the similarity of scores made using different methods—called convergence—is used as indirect evidence of construct validity.

11.2.3 Level of Measurement

Measurement scales are classified as nominal, ordinal, interval, and ratio. Table 11-1 describes each kind of measurement classification, lists the mathematical operations that can be performed on data from scales that achieve the given level of measurement, and gives an example of a scale of each type. The mathematical and statistical operations required in decision analysis are addition, subtraction, multiplication, and division. To be useful, a scale measuring preferences for health states must achieve measurement on at least interval scale, because the mathematical operations required in decision analysis are valid only on scales that achieve this level of measurement.

11.2.4 Strategies for Developing Scales

Two different overall strategies for developing scales that measure preference for health states have been described: holistic strategies and decomposed strategies (Froberg and Kane 1989B), which are also called multiattribute strategies (Pliskin, Shepard, Weinstein 1980). Holistic strategies require raters to assign scale values to each possible health state of interest. Decomposed, or multiattribute, strategies

Table 11-1 Measurement scales, permissible mathematical operations, and an example of each scale

Measurement Achieved by Scale	Description	Permissible Operations	Example
Nominal	States are assigned to categories without numerical meaning	None	Religion: Baptist, Catholic, Jewish
Ordinal	States are rank-ordered, but the distance between ranks does not have a numerical interpretation	None	Health status: excellent, good, fair, poor
Interval	States are rank-ordered, and the distance between ranks provides some information on the amount of difference between ranks	Addition, subtraction, multiplication, division	Beck depression scale
Ratio	States are rank-ordered, and the distance between different points on the scale provides information on the amount of difference between ranks	Addition, subtraction, multiplication, division, invariance with multiplication by a constant	Temperature

use statistical methods to develop a rating scale based on the rating of single attributes of health states or selected combinations of all possible health states.

Holistic strategies can place a heavy response burden on subjects if the number of health states to be rated is large. Decomposed strategies enable the investigator to obtain values for all health states without requiring the raters to assign values

Table 11-2 Descriptions of health status describing patients with cancer

Case 1: I am in the age range 40–64 years. I am unable to work. I am tired and sleep poorly due to discomfort in my back and arm. I am worried about my health and finances. I am able to drive my car and I make an effort to walk about my neighborhood.

Case 2: I am in the age range 40–64 years. Although I worked until recently, I am presently hospitalized and on complete bed rest. A nurse bathes, dresses, and feeds me. I am confused about time and place and have some memory loss. I have vomited and am only able to take clear fluids by mouth. I am dehydrated (lack water) and receive fluids by vein (IV therapy). I am incontinent (unable to control my bowels and bladder). I have low back pain.

Case 3: I am in the age range 40–64 years. I have been tired and weak and unable to work. I have lost 15 pounds in weight. I walk slowly, and travel outside the house is difficult. Much of the day I am alone, lying down in my bedroom. Social contact with my friends is reduced.

Case 4: I am in the age range 40–64 years. I am able to work. Over the past year I have noticed a feeling of tiredness and I have lost 20 pounds in weight. I have little energy and I am unable to keep up with my usual routine. I have made an effort to walk to work but I have let the house and hobbies "slide." I am sleeping poorly. I am maintaining my present weight.

Source: Llewellyn-Thomas et al. (1984).

to every one. The statistical methods used to derive rating scales from measurements made on decomposed health states are complex, and methods for multiattribute scaling are not considered further in this book. The interested reader is referred to Torrance, Boyle, and Horwood (1982) and to Boyle and Torrance (1984).

11.3 DEVELOPING MEASUREMENT SCALES IN PRACTICE

11.3.1 Overview

Developing a scale to measure preferences for health states involves several discrete steps (Torrance 1982; Froberg and Kane 1989A). First, the health states of interest are defined and described. Second, a rater or group of raters provides information on their preferences for each of the health states. Last, the information provided by the raters is used to create the numerical scale.

11.3.2 Defining and Describing the Health States

The health states selected will depend on the topic of the analysis. Usually a health state is defined for every possible distinct outcome of the intervention and its alternatives (Torrance 1982).

> *EXAMPLES:* In a comparison of different treatments for babies undergoing care in an intensive care unit, the health states that should be defined are all of the possible outcomes of the treatments and of the lack of treatment for the newborn.
>
> The health states that should be defined in a comparison of treatments for cancer should include all of the outcomes of having the cancer and not being treated and all of the outcomes of the treatments for cancer.

After the health states have been defined, they must be described. The descriptions of the health states should be put in functional or behavioral terms (Torrance 1982). Each description should include a statement on the level of physical health, emotional health, and everyday function in social and role activities, and general perceptions of well-being (Ware 1987). Psychological studies suggest that only five to nine separate items can be considered simultaneously (Miller 1956), and each health state description should contain no more than five to nine separate aspects (Torrance 1982; Froberg and Kane 1989A).

> *EXAMPLE:* Table 11-2 presents narrative descriptions of some of the health states describing patients with cancer that Llewellyn-Thomas et al. (1984) used in a study that compared different ways of describing health states. Each description is in functional and behavioral terms, and each one deals with all of the areas described above. There are from seven to nine separate statements about function within each description, and none contains more than nine.

11.3.3   Choosing Raters

Ideally, the raters should be selected according to the purpose of the study (Torrance 1982). For example, when the conclusion of the analysis applies to the society as whole, a representative sample should be used; when it applies to patients with a given disease, patients should be used. In practice, it is common to use convenience samples of patients, students, or professionals as raters. Attempts to justify this practice are based on the argument that preferences are stable across groups. This is not an adequate justification.

The published literature on measurement of preferences provides little guidance on the choice of the number of raters to use in a study to measure preferences for health states. It is common to encounter scales of health preferences developed based on the responses of only a handful of raters. It is difficult to apply formulas for formal sample size estimation to the problem of deciding how many raters to use, because a measure of the variation in the scale rating is unavailable. The number of raters used a study of health preferences should be chosen based on consideration of the variability of the measure, perhaps as measured in a pilot study or in studies by others.

11.3.4   Rating the Health States to Create the Numerical Scale

11.3.4.1   Overview

After the health states have been defined and described and the raters chosen, the health states must be rated. Rating methods fall into three categories: the standard gamble method, the time trade-off method, and direct scaling methods. There is also a set of miscellaneous methods that do not yield numerical scale data, which are described here for completeness.

The standard gamble is conceptually linked with decision theory but difficult to implement in practice. The time trade-off method is intuitively appealing and fairly simple to carry out, and it is used with increasing frequency in decision analysis. Direct scaling methods are the most common methods used to create scales, mostly because they are simple for the investigator to develop and for the subjects to understand. Other methods—the willingness-to-pay method and the equivalence method—are of interest because of they have been used often in cost-effectiveness analysis.

11.3.4.2   Standard Gamble

The standard gamble is a method for measuring preferences for health states that is directly derived from decision theory. It is considered the "criterion" method for obtaining information on preferences (Llewellyn-Thomas et al. 1984), as it incorporates the conceptual framework for decision making under conditions of uncertainty that is at the heart of decision theory.

Application of this method involves having raters choose between two alternatives. One alternative has a certain outcome and one alternative involves a gamble. The certain outcome is the health state to be rated. The gamble has two pos-

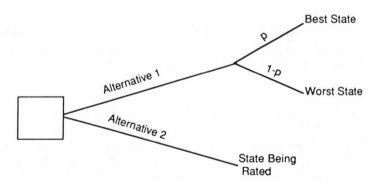

Figure 11-1   Graphical representation of standard gamble.

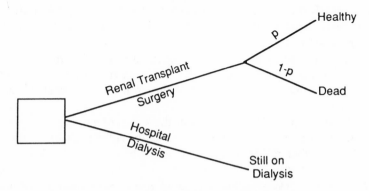

Figure 11-2   Graphic representation of standard gamble for the choice between renal transplant and continued hospital dialysis.

sible outcomes—the best health state (usually complete health), which is described as occurring with a probability p; or an alternative state, the worst state (usually death), which is described as occurring with a probability of $1 - p$. The probability p is varied until the rater is indifferent between the two alternatives. That is, the rater is indifferent between the alternative that is certain and the gamble that might bring the best health state.

EXAMPLE: Figure 11-1 graphically depicts a hypothetical gamble. Figure 11-2 depicts a gamble where the alternatives are renal transplant surgery and continued hospital dialysis. In the gamble, continued hospital dialysis is the health state that is stated to have a certain outcome, which involves continued—poor health and impaired social and role function. Renal transplant is the gamble: it has a probability p of bringing about a state of complete health, but it also is associated with a probability $1 - p$ of death due to surgery. A rater is asked to state at what probability p he or she would be willing to accept the certainty of the state of poor health and impaired social and role function instead of taking the risk of death from surgery as a means of possibly achieving a state of complete health. When p is close to 1.0, meaning that the probability of achieving complete health by gambling on surgery is close to 1, most raters would choose renal transplant surgery. When p is low, meaning that choosing the gamble has a high probability of causing death due to surgery, most raters would choose the certainty of the state of poor health and impaired social and role function.

The gamble is repeated for all of the health states that are to be rated. The points of indifference associated with each health state are the values used in the scale of health preferences.

Torrance (1987) described visual aids that help in the development of a rating scale using the gamble. A probability wheel, or "chance board," consisting of a disk with two movable, different-colored sections, is presented. The alternatives are displayed to the rater on cards, with the two outcomes of the gamble alternative color-coded to the two sectors of the probability wheel. The rater is told that the chance of each outcome is proportional to the similarly colored area of the disk. The rater is asked to adjust the sections to represent the preference for the gamble alternatives.

Even with visual aids, it is difficult for some raters to think in terms of probabilities. The number of health states that must be rated to derive a scale is often large, and resource demands and response burden limit the use of the standard gamble in practice.

The reliability of scales developed from the standard gamble has been evaluated infrequently. Table 11-3 summarizes evaluations of the reliability of the standard gamble and other rating methods. When it has been evaluated, intrarater reliability of the standard gamble has been good (Torrance 1976; Froberg and Kane 1989B). Test–retest reliability was good at one week but poor at one year (Froberg and Kane 1989B).

The ability of scales based on the standard gamble to achieve measurement on an interval or ratio scale has been assumed rather than proven. The standard gamble is considered to be the "criterion" method, and criterion validity has not been evaluated. There is a correlation between measures of preference based on the standard gamble and measures based on the time trade-off and category rating methods ($r = 0.56$–0.65) (Read et al. 1984), but the standard gamble produces consistently higher preference scores. Llewellyn-Thomas et al. (1982) found that small changes in the description of the gamble outcome can affect the value of

Table 11-3   Reliability of methods for measuring preferences for health states

Measure of Reliability	Method			
	Standard Gamble	Time Trade-off	Rating Scale	Willingness to Pay
Intrarater reliability	0.77	0.77–0.88	0.70–0.94	—
Test–retest reliability				
One week or less	0.80	0.87	0.77	—
One year	0.53	0.62	0.49	0.25
Interrater reliability	—	—	0.75–0.77	—

Source: Froberg and Kane (1989B).

preferences for health states provided by the same raters, showing internal inconsistency of utility measures based on the standard gamble.

11.3.4.3   Time Trade-Off

The time trade-off method was developed as an alternative to the standard gamble by Torrance, Thomas, and Sackett (1972). The time trade-off method also presents the rater with a choice. The choice is between two alternatives that both have a certain outcome. The choice is how many years of life the rater would be willing to give up to be in the healthier state compared with the less healthy one. Torrance (1987) has also described a useful visual aid for developing scales based on the time trade-off method. It is not described here.

The time trade-off method is easier for raters than the standard gamble. The response burden is about the same as for the standard gamble, but the task is more intuitive both for the rater and for the investigator.

Scales derived from the time trade-off method are not routinely subjected to evaluations to determine their reliability. Intrarater reliability was as good or better than the standard gamble when evaluated by the same investigator (Torrance 1976). As shown in Table 11-3, similar to the standard gamble, test–retest reliability has been found to be good at one week or less and poor at one year (Froberg and Kane 1989B).

Whether or not scales derived from the time trade-off method achieve measurement on a ratio or interval scale is unknown. Scales based on the time trade-off method are moderately well-correlated ($r = 0.56\text{–}0.65$) with scales based on the standard gamble and category rating methods (Read et al. 1984). The high correlation among scales does not prove that any of the scales measures what it purports to measure. It means that they are the same, not that they are necessarily valid (Read et al. 1984).

11.3.4.4   Direct Rating Scales

Methods for directly constructing a rating scale are the most commonly used methods for measuring preferences for health states. These methods include interval scaling, category rating, and magnitude estimation (Froberg and Kane 1989B).

Interval scaling starts by depicting the scale as a line on a page with clearly defined endpoints, called anchors. The rater identifies the best and worst health states and places these at the anchor points. The rater then rates the preference for each health state by placing each state at a point on the line between the anchors.

In category rating, raters sort the health states into a specified number of categories, and equal changes in preference between adjacent categories are assumed to exist. In magnitude estimation, the rater is given a "standard" health state and asked to indicate, with a number or a ratio, how much better or worse each health state is compared with the standard.

Visual aids—"feeling thermometers," boxes, decks of cards—are useful (Torrance 1987). These methods are the simplest to apply.

Table 11-3 shows that rating scales have good intrarater reliability ($r = 0.70\text{–}$

0.94) and interrater reliability ($r = 0.75–0.77$) when evaluated. Test–retest reliability is worse than for rating scales developed using the standard gamble and time trade-off methods both in the short term and in the long term.

The level of measurement of scales developed using these methods is assumed to be interval or ratio, but this assumption is unproven. Construct validity has not been evaluated. As noted previously, the scales based on these methods are moderately well correlated with scales based on the other methods ($r = 0.56–0.65$), but this does not assure validity.

11.3.4.5 Miscellaneous

The equivalence method asks the rater to state how many people in one health state are equivalent to a specified number of people in another health state. The willingness-to-pay method asks raters to state what percentage of the family income they would be willing to pay in order to be free of the stated condition. These methods have been used in cost-effectiveness analysis, but they are mostly not evaluated. They have not been used much in decision analysis or cost-effectiveness analysis in medicine.

11.4 INCORPORATING MEASURES OF PREFERENCE FOR HEALTH STATES INTO DECISION ANALYSIS

Notwithstanding the formidable difficulties of constructing scales that measure health preferences and theoretical concerns about their use, these scales are used with increasing frequency in decision analysis and in cost-effectiveness analysis. They are most often incorporated into the decision analysis by using the scale measure to "adjust" the outcome derived from the decision analysis in order to obtain a measure of quality-adjusted life years. Quality-adjusted life years is a measure that attempts to combine expected survival with expected quality of life in a single metric (LaPuma and Lawlor 1990). An analysis that uses quality-adjusted life years seeks to evaluate the trade-off between mortality, morbidity, the preferences of patients and society for various types of morbidity, and the willingness of patients and society to accept a shortening of life to avoid certain morbidities. The concept of the quality-adjusted life year is explicit recognition that in order to achieve certain states of health, people are willing to take a measurable risk of a bad outcome.

> *EXAMPLE:* Angina pectoris has been shown in several randomized clinical trials to be more common in patients who are managed medically than in patients who undergo coronary artery bypass graft (CABG), even though life expectancy is not shorter in patients managed medically. Weinstein and Stason (1985) did an analysis of CABG surgery that attempted to incorporate information about the effects of coronary artery surgery on relief from angina into estimates of the effect of CABG on life expectancy. The analysis adjusted for the diminished quality of life that accompanies angina pectoris. Weinstein and Stason used the time trade-off method as a

measure of the preferences for health. In the analysis, a patient with severe angina pectoris was assumed to be willing to accept 7 years of life with no symptoms in exchange for 10 years of life with severe angina.

The value assigned to each health state is often called Q.[1] The Q factor for each health state is the number that is used to adjust the life expectancy for each health state. Life expectancy estimated by the methods described in Chapter 9 is multiplied by Q to estimate the equivalent number of quality-adjusted life years.

EXAMPLE: In the analysis by Weinstein and Stason (1985), for the patient with severe angina pectoris,

$$Q = \frac{7}{10} = 0.7$$

A patient with mild angina was assumed to be willing to accept 9 years of life with no symptoms in exchange for 10 years of life with mild angina. For patients with mild angina

$$Q = \frac{9}{10} = 0.9$$

In practice, a decision analysis that estimates quality-adjusted life years proceeds as described in Chapters 9 and 10 up to the final calculation of expected outcome. In a utility analysis, the number of years spent in a given health state is multiplied by the corresponding value of the scale measuring the preference for that health state, Q, yielding an estimate of the number of years of life with full health that is equivalent to the number of years in the given state of morbidity (Weinstein and Stason 1977). The adjusted measures are then summed and compared.[2]

EXAMPLE: Chapter 9 described a decision analysis that estimated life expectancy in patients with bleeding esophageal varices who did and did not undergo sclerotherapy. Life expectancy in the no-sclerotherapy group was estimated to be 1.87 years, and in the sclerotherapy group it was estimated

Table 11-4 Incorporation of quality adjustment into decision analysis comparing sclerotherapy with no sclerotherapy

	No Sclerotherapy			Sclerotherapy		
	Well	Bleed	Diminished Health	Well	Bleed	Diminished Health
Number of cycles	62,495	124,996	62,498[a]	86,202	125,860	62,930[a]
Quality-adjusted life expectancy: no sclerotherapy					$0.62 + 0.62 = 1.24$ years	
Quality-adjusted life expectancy: sclerotherapy					$0.86 + 0.63 = 1.49$ years	
Difference					$1.49 - 1.24 = 0.25$ year	

[a] After adjustment for diminished quality of life during years where a bleed occurs; $Q = 0.5$.

to be 2.12. The expected effect of sclerotherapy compared with no sclerotherapy is to increase life expectancy by 0.25 year. Years of life when bleeding occurs are not equivalent to years of complete wellness. They are years of diminished health. Imagine that these years are rated as having a value Q of 0.5 relative to complete wellness. To take into account the lower utility of years of life after an initial bleed, the years of life in the states labeled "bleed" in Table 9-4 are multiplied by Q before they are summed. The average cycles based on the quality-adjusted figures are used to estimated quality-adjusted life expectancy, as shown in Table 11-4. For no sclerotherapy, quality-adjusted life expectancy is 1.24 years. For sclerotherapy, quality-adjusted life expectancy is 1.49 years. The benefit of sclerotherapy compared with no sclerotherapy is to add 0.25 quality-adjusted life year. This number can be compared with the number of life years gained with sclerotherapy calculated in Chapter 9—that is, 0.25 year. In this example, the adjustment did not alter the conclusion, although the analysis has incorporated diminished quality of life and may be preferred for this reason.

11.5 LIMITATIONS OF MEASURES OF PREFERENCES FOR HEALTH STATES

The validity of incorporation of measures of preferences for health states into decision analysis rests on the assumption that preferences are well defined, consistent, and quantifiable. There are some concerns about the degree to which currently available measures meet these criteria.

The measurement of preferences for health states has been shown to be influenced by the measurement process itself, including the way the information is presented, as well as by the method used to elicit ratings and other circumstances of the rating process (Llewellyn-Thomas et al. 1984). The different scaling methods do not yield identical results even when the same problem is presented (Llewellyn-Thomas et al. 1984, Read et al. 1984), and there is no consensus about which method is the best method. It is also uncertain what population should be used to represent societal preferences—patients, the general population, or a panel of experts. On the conceptual level, Llewellyn-Thomas et al. (1984) have suggested that it might be naive to think of any state of health as possessing a single value for the whole population.

Additional moral and ethical concerns about the use the quality-adjusted life year to set public policy are described in Chapter 15.

NOTES

1. This Q should not be confused with the statistic called Q used to test homogeneity of effect size that was described in Chapters 7 and 8.

2. The values of the adjusted measures are discounted when appropriate. Discounting of measures of benefit is covered in Chapter 12.

12

Advanced Cost-Effectiveness Analysis

Cost-effectiveness analysis begins with a decision analysis that compares the effectiveness of alternative courses of action. Once the decision analysis has been carried out, costs are measured and the cost effectiveness of the alternatives is determined. This simplistic description of cost-effectiveness analysis belies extraordinary complexity in the definition of the concept of cost effectiveness and in the estimation of cost. Handling of costs and benefits that occur in the future also poses analytic challenges.

Section 12.1 presents key concepts in cost-effectiveness analysis including descriptions of the various definitions of cost effective, the distinction between average and marginal cost-effectiveness ratio, the concept of the perspective of the analysis, as well as the basic mathematical formulation of cost-effectiveness analysis. Section 12.2 presents the conceptual basis for estimating cost and Section 12.3 describes how to estimate cost in practice. Sections 12.4 and 12.5 discuss two important issues related to analysis of cost—discounting of cost and inflation. Section 12.6 discusses discounting of benefits. Section 12.7 discusses time horizon in cost-effectiveness analysis.

12.1 KEY CONCEPTS

12.1.1 Definitions of Cost Effective

In medical applications, the term cost effective is used in many ways (Doubilet, Weinstein, McNeil 1986). Because cost-effectiveness analysis has as an implicit

Table 12-1   Ways that the term "cost-effective" is misused in medical applications

In the absence of data on both cost and effectiveness

When effectiveness is demonstrated, in the absence of data on cost

In the restricted circumstance where the intervention is cost saving relative to alternatives

Source: Doubilet, Weinstein, McNeil (1986).

goal the determination of whether a treatment or intervention is or is not cost effective, an understanding of the various uses and misuses of the term is important. Agreement on how the term should be used is a necessary prelude to meaningful interpretation of claims that an intervention is cost effective.

Doubilet, Weinstein, and McNeil (1986) describe three ways that the term "cost effective" is misused in medical applications. These misuses are listed in Table 12-1. First, an intervention is sometimes called cost effective in the absence of data on both cost and effectiveness. The term cost effective is meaningless without information both on cost and on effectiveness. The term is also misused as a synonym for effectiveness in the absence of information on cost. Explicit information on cost is a necessary condition for assessing and claiming that an intervention is cost effective. The term cost effective is sometimes restricted to situations where the intervention is cost saving relative to its alternatives. This restriction stems from the origin of cost-effectiveness analysis in business economics. In business, the ability to reduce everything to a monetary value is easier than it is in medicine, and cost effectiveness can be based on strict comparison of monetary cost. Restricting the term cost effective to situations where an intervention is cost saving is considered too narrow a definition of cost effective in medical applications (Doubilet, Weinstein, McNeil 1986).

Doubilet, Weinstein, and McNeil (1986) suggest that in medical applications the term cost effective should be used when an intervention *provides a benefit at an acceptable cost.* Within this framework, they identify four criteria for calling an intervention cost effective. These criteria are listed in Table 12-2. First, an intervention is cost effective when it is less costly and at least as effective as its alternative. Second, an intervention is cost effective when it is more effective and more costly, but the added benefit is worth the added cost. Third, an intervention is cost effective when it is less effective and less costly, and the added benefit of the alternative is not worth the added cost. Last, an intervention is cost effective when it is cost saving, and the outcome is equal to or better than the alternative.

Doubilet, Weinstein, and McNeil (1986) recognize that interventions that are

Table 12-2   Recommended criteria for calling an intervention "cost-effective" in medical applications

Less costly and at least as effective

More effective and more costly, with the added benefit worth the added cost

Less effective and less costly, with the added benefit of the alternative not worth the added cost

Cost saving with an equal or better outcome

cost saving fall within the broad framework for calling an intervention cost effective, but they suggest that these interventions be called "cost saving" rather than cost effective. Other authors (Warner and Luce 1982) consider that interventions that are cost saving should also be called cost effective; this view is the more widely accepted view and will be adopted here.

Accepting these criteria as the criteria for calling an intervention cost effective does not resolve the ethical dilemmas of allocating scarce resources. Except when an intervention is cost saving, the criteria do not avoid difficult value judgments. Thus, "worth the added cost" is not an economic issue, but an ethical and moral issue. Opinions about whether or not something is "worth" a certain amount of money are subject to variations in the perspective and value structure of the person making the judgment of worth. What is an acceptable cost in one setting may be unacceptable in another. What is acceptable at one time may not be acceptable at another. The limitations of cost-effectiveness analysis as a way to make decisions about allocation of resources in situations of scarcity are considered in more detail in Chapter 15.

12.1.2 Average Versus Incremental Cost Effectiveness

Cost effectiveness is measured as a ratio of cost to effectiveness. A distinction is made between an *average* cost-effectiveness ratio and an *incremental* or *marginal* cost-effectiveness ratio (Detsky and Naglie 1990). An average cost-effectiveness ratio is estimated by dividing the cost of the intervention by a measure of effectiveness without regard to its alternatives. An incremental or marginal cost-effectiveness ratio is an estimate of the cost per unit of effectiveness of switching from one treatment to another, or the cost of using one treatment in preference to another. In estimating an incremental or marginal cost-effectiveness ratio, the numerator and denominator of the ratio both represent differences between the alternative treatments (Weinstein and Stason 1977):

$$\text{incremental cost effectiveness} = \frac{\text{difference in cost}}{\text{difference in effectiveness}}$$

where

$$\text{difference in cost} = \text{cost of intervention} - \text{cost of alternative}$$

and

difference in effectiveness

$$= \text{effectiveness of intervention} - \text{effectiveness of alternative}$$

The average cost-effectiveness ratio and the incremental or marginal cost-effectiveness ratio are identical only in the highly unusual situation where the alternative treatment has a zero cost and no effectiveness (Detsky and Naglie 1990).

> *EXAMPLE:* Bone marrow transplantation for acute nonlymphocytic leukemia costs $193,000, and its average effect in patients with the disease is to add 3.32 quality-adjusted life years (Welch and Larson 1989). Using this information, it is possible to estimate the average cost-effectiveness ratio

of bone marrow transplantation as \$193,000/3.32, or \$58,132 per quality-adjusted life year. Imagine for the purpose of illustration that failure to do bone marrow transplantation in a patient with acute nonlymphocytic leukemia always results in death and that patients who do not undergo the procedure incur no costs because the disease is very rapidly fatal. The incremental cost effectiveness of bone marrow transplantation compared with the alternative, doing nothing, is also \$58,132 since

$$\frac{\text{cost difference}}{\text{effectiveness difference}} = \frac{\$193,000 - 0}{3.32 \text{ QALY} - 0 \text{ QALY}}$$

Proper cost-effectiveness analysis is always comparative. That is, the average cost-effectiveness ratio is not a useful quantity (Detsky and Naglie 1990). In many cases, the implicit alternative to an intervention is doing nothing. But doing nothing usually has costs and effects that should be taken into account in the cost-effectiveness analysis (Detsky and Naglie 1990). Average cost-effectiveness ratios will be appropriately estimated when decision analysis is the first step in the cost-effectiveness analysis, since decision analysis always compares an intervention with at least one alternative.

12.2   COSTS

12.2.1   Terminology

The terminology of cost-effectiveness analysis is sometimes confusing. There appears to be consistency in use of certain terms in textbooks of economics, but the standard economic terminology often becomes garbled in medical applications. The term "indirect" cost is sometimes used by medical authors (Weinstein and Fineberg 1980) to refer to the component of production costs that includes rent, maintenance, and utilities; here this is called "overhead." Production costs are sometimes called resource costs. What are called here "indirect" costs are called opportunity costs by others (Sox et al. 1988). Table 12-3 gives the definitions of key terms as they are used here and the terms that are used by other authors for the same concept.

12.2.2   Framework

The economic concept of opportunity cost is central to cost-effectiveness analysis. The opportunity cost of a resource is its total value in another use. Opportunity cost consists of production costs and indirect costs. Production costs are the actual amount of resources consumed in the production of the good or the provision of service. Indirect costs are the monetary values of other resources in their alternative use.

EXAMPLE: A man goes to the hospital to get a chest x-ray. To get the chest x-ray, he takes an hour off from his regular job. The total opportunity cost of the visit for the x-ray includes the production cost of the x-ray and

Table 12-3 Definitions of key terms describing components of cost and terms used by others for the same concept

Term Used Here	Definition	Term Used by Others for the Same Concept
Production cost	The total amount of resources consumed in the production of a good or a service	Resource cost
Direct cost	The cost of the materials and labor that go directly into production of a good or service, such as labor, materials, and equipment	Same
Overhead	The cost of all other items needed to produce a good or service, such as space, energy, and administrative services	Indirect cost
Induced cost	Costs, positive or negative, that are caused by an intervention and that would not have been incurred in the absence of the intervention, such as the costs of treating side effects of a medication and the cost of continuing to treat persons who live longer because of an intervention	Same
Indirect cost	The cost of lost productivity and monetary values, both positive and negative, associated with the alternative use of time; other costs that are neither production cost nor overhead	Opportunity cost
Opportunity cost	The value of all costs in an alternative use	Total cost

the indirect cost. The production cost of the x-ray consists of all the resources that go into provision of the service, which include the cost of labor to make the x-ray and read it, the cost of the x-ray film, and the overhead cost, which includes rent on the building in which the service is delivered and a number of other items. The indirect cost consists of the amount of lost wages for the hour that the man took to get the x-ray.

12.2.3 Perspective and the Importance of Perspective

Costs are seen differently from different points of view. For example, the cost of hospitalization from the perspective of an insurance company is the amount of money that the company pays to the hospital for that illness under the coverage plan for the individual who is hospitalized. The cost from the perspective of the hospital is the true cost of providing the service, which includes the labor costs, the costs of the building in which the services are provided, and other overhead costs. The cost of the hospitalization from the perspective of the person who is hospitalized is the amount paid out-of-pocket, which is often the amount not covered by insurance. The cost of the hospitalization from the perspective of society includes the total cost to society of providing the service and of lost productivity due to the illness as well as any benefits that arise from future gains in productivity and aversion of costs in the future.

It is important to state explicitly the perspective of a cost-effectiveness analysis, since the perspective determines which costs should be included in the analysis and what economic outcomes are considered benefits. Usual perspectives in cost-effectiveness analysis are the societal perspective and the program perspective.

An analysis that takes the societal perspective seeks to determine the total costs of the treatment or intervention to all payers for all persons. Taking the societal perspective, the cost of the intervention is offset by reductions in lifetime medical care even if these occur long after the intervention. Taking the societal perspective, the cost of the intervention may also be considered to be offset by increased economic productivity due to enhanced life expectancy.

Analyses that take the program perspective are more heterogeneous in their aims. An analysis that takes the program perspective might, for example, address the question of the immediate cost of a treatment and its outcome in order to compare it with other treatments and outcomes for the same condition. It might seek to determine whether coverage for the treatment would save money for the program in the long run. For programs where coverage is not lifelong, the cost of a treatment or intervention is not offset by reductions in medical care costs that occur after coverage ends. Savings in medical costs not covered by the program also are not considered to offset the cost of the intervention in analyses that take the program perspective.

12.2.4   Production Cost

Weinstein and Fineberg (1980) present a breakdown of the components of production cost for a medical cost-effectiveness analysis (Table 12-4). The first general category of production costs includes the direct costs: cost of equipment, labor, and materials. The second general category of production cost is overhead, con-

Table 12-4   Components of production cost

Direct cost
Equipment
Labor: professional and nonprofessional
Materials
Overhead
Rent/building depreciation
Space preparation
Maintenance
Utilities
Support services
Other administrative services
Induced costs (and savings)
Treatments added or averted
Tests added or averted

Source: Weinstein and Fineberg (1980).

sisting of rent and depreciation, space preparation and maintenance, utilities, other support services, and administrative support services. The last general category of production cost is induced costs. Induced costs include costs due to added or averted treatments or tests. That is, savings due to averted treatments or tests that are attributable to the intervention are considered to be negative costs, not benefits, in cost-effectiveness analysis.

12.2.5 Indirect Cost

Indirect cost includes the costs of lost wages and productivity and other values of the resources and time. Indirect costs may be negative costs if the treatment or intervention leads to gains in productivity over the short term or the long term.

Indirect costs include not just the monetary value of the use of the patient's time in other economically productive activities, but also the monetary value of other uses of resources and other values of time. For example, imagine that the man who goes to get an x-ray is retired. The one-hour visit to get the x-ray does not result in lost wages or lost productivity at work. The time has value to the man, however. He might rather have been playing golf or watching television. The time may even have monetary value to the man; he might be willing to pay $40 *not* to go to the doctor. These costs are indirect costs.

Fear, inconvenience, and pain caused by a medical treatment are also indirect costs of the treatment. It is difficult to incorporate these costs into the analysis as monetary costs. Cost-utility analysis, in which the measure of outcome is a utility measure that takes into account the effect of the treatments on quality of life, is the most common way that these factors are incorporated into a cost-effectiveness analysis.

12.3 ESTIMATING COSTS

12.3.1 Direct Measurement of Production Cost

Having identified the components of the production cost for a medical good or service, the difficult problem of estimating each component remains. Table 12-5 shows Weinstein and Fineberg's (1980) suggestions for estimating each of the components of the production cost of an x-ray procedure. The reader is referred to Weinstein and Fineberg's text for more information on the direct measurement of production cost. The text of Sugden and Williams (1990) also discusses direct measurement of production cost, and the interested reader is referred also to this book.

12.3.2 Use of Charge or Payment to Estimate Production Cost

12.3.2.1 Overview

In practice, production cost is rarely measured directly as the amount of resources needed to produce a good or a service, because obtaining this information requires

Table 12-5 Components of production costs and basis for estimating these costs for a cost-effectiveness analysis of an x-ray procedure

Component	Basis for Estimation
Direct cost	
Equipment	Depreciated cost divided by frequency of use
Labor	
Radiologist's time	Professional fee
Time of other personnel	Hours times wages per hour
Materials	
X-ray film	Cost per procedure
Other supplies	Cost per procedure
Overhead	Fraction of other hospital costs allocated to this procedure

Source: Weinstein and Fineberg (1980).

studies that involve primary data collection and considerable expertise in accounting. In spite of well-reasoned arguments against indiscriminate use of charge and payment data as a substitute for cost measured correctly in economic terms (Finkler 1982), charge or payment data are almost always used in medical cost-effectiveness analysis.

Charge is the amount the hospital, the clinic, the physician, or the pharmacy attempts to recover (or bills) for a good or service. Payment is the amount actually paid by the individual or the third-party payer for the good or service.

Using charge or payment may be correct when the perspective of the analysis is a program perspective, since payments and charges are real costs from the point of view of the program. When the perspective of the cost-effectiveness analysis is societal, information on payment and charge as a substitute for cost can lead to unwarranted conclusions (Finkler 1982). The limitations on interpretation of cost-effectiveness analysis that result from the use of charge and payment as a substitute for cost are discussed further in Chapter 15.

Estimating charge or payment is not a trivial task. There are few centralized sources of cost or payment data for nationally representative samples of the population. Charges and payments may vary from place to place for the same intervention or between patients for the same condition, and they may change rapidly over time.

12.3.2.2 Use of Medicare Data and Information from Insurance Databases

The schedule for Medicare reimbursement for hospitalization may be used to estimate hospital costs. Hospital discharge diagnoses are reimbursed according the Diagnosis Related Group (DRG). Information on the amount that Medicare reimburses for each DRG is published and is reasonably accessible.

Use of DRG payments has some problems. First, most conditions have more than one associated DRG. For example, there are several DRGs for acute myocardial infarction, with different amounts of reimbursement depending on whether the myocardial infarction was or was not complicated. To use the cost of

Medicare reimbursement for myocardial infarction in a cost-effectiveness analysis requires some knowledge of the proportion of persons with myocardial infarction in the several DRGs for myocardial infarction. This information may be difficult to obtain. Second, the Medicare hospital payments for a given condition vary according to a number of factors, such as location and teaching status. For any given single DRG, the Medicare payment may differ by a large amount depending on these factors. A choice among the different DRG payments is difficult to make, since there is no fixed rule for making this choice.

One approach is to use the standardized payment for urban hospitals and to adjust the standardized payment to include the average allowance for outliers, disproportionate share, indirect teaching payments, and capital cost (Tosteson et al. 1990). The calculations necessary to make this adjustment are not given here.

Payments by Medicare and private insurance companies for hospitalization do not include payments to surgeons, radiologists, or anesthesiologists, or to physicians who attend the patient periodically during hospitalization or as consultants. These payments must be estimated separately and added the payment for hospitalization. Average fees for surgeons, radiologists, and anesthesiologists are usually estimated based on locally available data on fees. Lump-sum fees for the surgeon, anesthesiologist, and radiologist are added to the estimated cost of hospitalization. An estimate of physician fees for daily attendance for hospitalization is usually made by multiplying the overall length of stay for the hospitalized condition by the average of physician per diem charges for a hospital visit.

When the condition being examined is not one that is likely to be covered by Medicare, such as childbirth, it is sometimes possible to obtain information on hospital payment from major insurers, such Blue Cross. Data from this source are not nationally representative, however, and the level of detail may not be sufficient for the cost-effectiveness analysis.

> *EXAMPLE:* Smoking during pregnancy increases the likelihood of being low birth weight because of preterm birth and because of growth retardation without any effect on gestational age. The consequences of these two outcomes are very different. Preterm low birth weight may result in long and costly hospitalization, whereas low birth weight at term affects length of hospital stay and resource use by a much smaller amount. In an analysis of the cost effectiveness of smoking cessation programs in pregnancy (Shipp et al. 1992), we found that Blue Cross and other insurance companies were unable to provide us with information on the cost of hospitalization of infants with these two outcomes separately.

12.3.2.3 Sample Surveys

To estimate charges for drugs, a survey of local area pharmacies can be done. Hospital cost information can be based on review of samples of patients with the condition of interest.

> *EXAMPLES:* In a study that assessed the cost effectiveness of various alternatives to admission to the coronary care unit in patients with chest

pain, Fineberg, Scadden, and Goldman (1984) reviewed the billing records of a random sample of 15 patients with chest pain who were admitted to one hospital in 1980 and were found not to have a myocardial infarction.

In our study of the cost effectiveness of smoking cessation programs in pregnancy (Shipp et al. 1992), we identified all of the infants born in two hospitals who were low birth weight. The infants were separately classified according to whether they were preterm or term. The hospital billing records of all of these infants and a random sample of infants of normal weight were reviewed to determine the average total charges to be used in the cost-effectiveness analysis.

Charge data based on a sample survey of patients hospitalized at one hospital may not be representative of charges for the same condition in other hospitals. Similarly, drug costs may vary by region, and a local area survey may not be representative. Unless the survey is large, the estimate of cost may be statistically unstable. Charges for hospitalization for many conditions tend to be highly skewed, and outliers may dominate the estimate of charge when the estimate is based on a simple mean.

EXAMPLE: In the cost-effectiveness analysis of bone marrow transplantation by Welch and Larson (1989), the costs of chemotherapy and bone marrow transplantation were based on information from actual patients with acute nonlymphocytic leukemia. Figure 12-1 shows these estimates. For both the chemotherapy and transplantation groups, the distribution of cost is skewed, and there is at least one extreme outlier of cost.

In sample surveys based on fewer than 100 or so observations, the median charge is probably a better measure of charge to use in the cost-effectiveness analysis than the mean. When data are skewed, it may be wise to exclude the lowest and the highest charge before computing the mean. A sensitivity analysis using the mean cost and then the median cost, with and without outliers, may be useful.

12.3.2.4   Indirect Costs

It is common to ignore costs due to lost productivity and lost wages in a cost-effectiveness analysis. However, when information on these costs is available, they should be incorporated into a cost-effectiveness analysis. When the analysis takes the societal perspective, gains in productivity should be included as negative indirect costs.

12.4   DISCOUNTING COSTS

12.4.1   Overview

It is a tenet of economics that the value of a dollar now is greater than the value of a dollar later. The majority of people will prefer to receive $100 today than

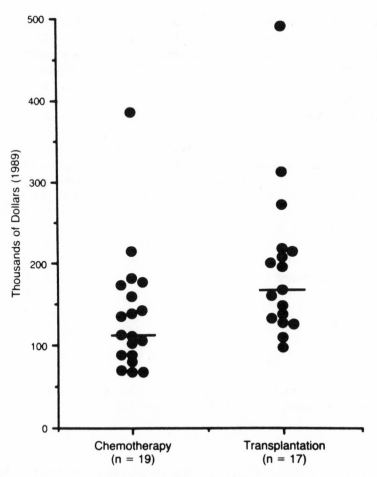

Figure 12-1 Five-year costs of chemotherapy and bone marrow transplantation for patients with acute nonlymphocytic leukemia as estimated from data on 19 patients who had chemotherapy and 17 patients who underwent bone marrow transplantation at one hospital. (Reproduced with permission from Welch and Larson, *New England Journal of Medicine*, 1989;321:810.)

$100 a year from today. This preference for a dollar today is logical because it would be possible to invest $100 received today in, for example, a savings account that pays 5% interest and receive $105 one year from today. The preference for a dollar today is called the *time preference for money.* In economic analysis, the time preference for money necessitates *discounting* future costs. Discounting adjusts future costs and expresses all costs and monetary benefits in terms of their present value.

In most medical situations, not all the costs of an intervention or treatment are incurred at a single point in time. For example, medical treatment of angina pectoris does not cure a patient of coronary heart disease, and so treatment is long term. The costs of treatment are incurred in every year of treatment. Similarly, the effects of a medical treatment or intervention often occur over a long period of time, and the long-range effects of the treatment or intervention may contribute to the total cost and the total cost savings of the treatment or intervention. For example, having coronary artery bypass surgery for an isolated lesion of the left main coronary artery at age 60 affects the likelihood of dying of an acute myocardial infarction in the year of surgery and for 5 to 10 years thereafter. The patient with this condition who undergoes coronary artery surgery is less likely to be hospitalized for an acute myocardial infarction in the years after surgery and will have lower costs of hospitalization for myocardial infarction over the next 10 years compared with a patient who elects medical management.

When costs of treatment or cost savings due to treatment occur over a long period of time, it is necessary to take into account the time preference for money by discounting future costs. The formula for discounting is a simple one:

$$c_{present} = c_0 + \frac{c_1}{(1 + r)^1} + \frac{c_2}{(1 + r)^2} + \cdots + \frac{c_n}{(1 + r)^n}$$

where $c_{present}$ is the cost in current dollars, r is the discount rate, and $c_0, c_1, \ldots, c_n$ are costs in future years.

EXAMPLE: At a discount rate of 5%, the cost, or present value, of a dollar spent 1 year from now is

$$\frac{\$1}{(1 + 0.05)^1} = \$0.95$$

At the same discount rate, 5%, the present value of a dollar spent 1 year from now and another dollar spent 2 years from now is

$$\frac{\$1}{(1.05)^1} + \frac{\$1}{(1.05)^2} = \$0.95 + \$0.91 = \$1.86$$

The process of discounting at a positive rate gives greater weight to costs and monetary benefits the earlier they occur. High positive discount rates favor alternatives with costs that occur late.

EXAMPLE: There is an operation for an otherwise fatal disease that costs $10,000 and cures all the patients with the disease. The alternative, a medical treatment, also cures patients with the disease, but the treatments must be given over 5 years. These treatments cost $2,000 per year. Without discounting, compared with doing nothing, the cost of surgery is $10,000 per life saved and the cost of medical treatment is $10,000 per life saved. Neither alternative is cost effective with respect to the other by any of the definitions of cost effective given earlier.

Assuming that the discount rate is 5%, the cost of surgery remains $10,000 per life saved. The cost of medical treatment after discounting is

$$\$2000 + \$1905 + \$1814 + \$1728 + 1645 = \$9092$$

After discounting, medical treatment is cost effective relative to surgery, because it is less costly with the same outcome. At any positive discount rate, medical treatment is cost effective relative to surgery.

12.4.2 Problems with Discounting

The preceding example illustrates two problems with discounting. Most important, extrapolation from this basic tenet of economic analysis to preferences for medical treatment may not be a good one. In the example, being cured now as compared with five years from now is a factor that probably would influence a choice between surgery and medical treatment. A preference for early cure is difficult to quantify in monetary terms, although measures of outcome adjusted for quality of life that were described in Chapter 11 could take it into account to some extent.

The second problem is that a conclusion about whether an intervention is cost effective relative to its alternative may be entirely dependent on discounting, as shown in the example. The dependence of conclusions on discounting tends to bother those not trained in economics, because the choice of the discount rate, discussed in the next section, is problematic.

12.4.3 Choice of the Discount Rate

In spite of the intuitive problems with discounting costs, economists seem sure that costs should be discounted (Drummond, Stoddart, Torrance 1987; Sugden and Williams 1990). Despite widespread consensus on the need for discounting, the appropriate discount rate remains a matter of controversy.

The discount rate is usually chosen as the rate of return on private investments, adjusted for inflation. Unfortunately, the rate of return on private investment is neither a fixed number nor invariate over time. The use of the private sector investment on return for public sector program costs, which comprise a large proportion of all medical costs, may not be correct (Sugden and Williams 1990).

Another approach to choosing a discount rate for cost-effectiveness analysis, suggested by several authors (Weinstein and Fineberg 1980; Warner and Luce 1982), is to pick a "reasonable" rate—Weinstein and Fineberg suggest 4–6%, Sugden and Williams (1990) suggest 3–5%—and test the sensitivity of the conclusions to assumptions about the discount rate, including a discount rate of zero. Table 12-6 shows that many recent published cost-effectiveness analyses have used a discount rate of 5%. A discount rate of 5% seems a reasonable rate to use in a cost-effectiveness analysis based on historic precedent.

Table 12-6 Discount rates used in some cost-effectiveness analyses published in 1987–1991

Study	Topic	Discount Rate (%)
Castellano and Nettleman (1991)	Prophylaxis for *Pneumocystis carinii* pneumonia	5
Tosteson et al. (1990)	Screening and hormone treatment for osteoporosis	5
Goldman et al. (1988)	Long term beta blockers after acute myocardial infarction	5
Kinosian and Eisenberg (1988)	Treatments for hypercholesterolemia	5

12.4.4 Effects of the Discount Rate on Estimated Cost Effectiveness

The choice of the discount rate affects the magnitude of the estimate of the cost effectiveness of an intervention compared with its alternatives.

EXAMPLE: Using a discount rate of 10%, the cost of surgery in the example given above is still $10,000 per life saved. The cost of medical treatment is $8340 per life saved. With a discount rate of 5%, surgery costs $10,000 per life saved, but the cost of medical treatment is $9092 per life saved.

Because the discount rate affects the absolute magnitude of the estimate of costs, it is essentially impossible to compare between interventions examined in different cost-effectiveness analyses done with different discount rates.

EXAMPLE: Consider two treatments, treatment A and treatment B. Each treatment costs $2000 per year for 5 years. The cost effectiveness of each treatment was compared with the doing nothing alternative by two different analysts who used two different rates to discount costs. Assume that both treatment A and treatment B cure all patients and have no other risks or benefits. As shown previously, with a discount rate of 10%, the cost of the treatment A is $8340 per life saved; with a discount rate of 5%, the cost treatment B is $9092 per life saved. Solely because of the difference in the choice of discount rates, treatment A, which costs $8340 per life saved, would be called cost effective because it is less costly and equally effective compared with treatment B. But the cost effectiveness of the treatment A is an artifact of the use of a higher discount rate.

12.5 INFLATION

Inflation is an issue that is separate from discounting. If the inflation rate for future medical costs is the same as the general rate of inflation for other goods and services, inflation can be ignored in cost-effectiveness analysis because the payment for inflated future costs will be in dollars that have inflated at the same rate. This can be shown mathematically (Sugden and Williams 1990) as follows.

Assume that the cost of some medical good or service is c_0 in year 0. If the inflation rate is i, then the cost of the medical good or service taking inflation into account is

$$\text{year 1: } c_1 = c_0 (1 + i)$$
$$\text{year 2: } c_2 = c_0 (1 + i)^2$$

The total is

$$c = c_1 + c_2 + \cdots + c_n$$

If the general inflation rate is also i and the discount rate is r, the total cost in current dollars is

$$c_{present} = c_0 + \left[\frac{c_0 (1 + i)}{(1 + r)(1 + i)} \right] + \left[\frac{c_0 (1 + i)^2}{(1 + r)^2 (1 + i)^2} \right]$$
$$+ \cdots + \left[\frac{c_0 (1 + i)^n}{(1 + r)^n (1 + i)^n} \right]$$

which simplifies to

$$c_{present} = c_0 + \frac{c_0}{(1 + r)^1} + \frac{c_0}{(1 + r)^2} + \cdots + \frac{c_0}{(1 + r)^n}$$

This amount, $c_{present}$, is the same as the discounted cost ignoring inflation that was shown in Section 12.4.2.

Inflation in medical care costs has not been the same as the general inflation rate for at least the last 10 to 15 years. The rate of inflation in hospital costs, for example, has greatly outstripped the general inflation rate. When the inflation rate for a medical care service is higher than the general inflation rate, taking the greater inflation rate of medical services into account in a cost-effectiveness analysis may alter conclusions based on it.

EXAMPLE: The health insurance arm of an insurance company will decide on the basis of cost-effectiveness analysis whether to pay $100,000 for bone marrow transplant for an advanced cancer or whether to pay for a medical treatment that costs $20,000 per year and lasts five years. The treatments are equally effective and, for the purposes of discussion, are assumed to have no other benefits or risks. Thus, the decision on payment is rationally made strictly on the basis of the cost effectiveness of transplantation relative to its alternative, medical treatment. Assume that the rate of return on investments for the insurance company as a whole is 10%. Using the preceding formulas and discounting at a rate of 10%, the cost of surgery is $100,000. The discounted cost of medical treatment is

$20,000 + $18,181 + $16,529 + $15,026 + $13,660 = $83,396

Based on this analysis, medical treatment is cost effective compared with transplantation, because it is less costly and equally effective.

If the inflation rate for medical services is assumed to be 20%, the cost of transplantation, which is a one-time cost, remains $100,000. The cost of medical treatment after taking inflation into account is

$20,000 + $24,000 + 28,800 + 34,560 + 41,472 = $148,832

Discounting these costs at 10% gives

$20,000 + $21,818 + 23,801 + $25,965 + $28,326 = $119,910

Under these assumptions, transplantation is cost effective compared with medical treatment because it is less costly and equally effective.

Discounting the inflated costs at 20% instead of 10% is equivalent to assuming that the general inflation rate and the rate of inflation of medical goods and services are the same; it would yield the same answer as when inflation is ignored.

In practice, differences between the rate of inflation in medical care costs and the general inflation rate are usually ignored in cost-effectiveness analysis. In ignoring inflation, the implicit assumption is that the inflation rate for future medical costs is the same as the general inflation rate. This assumption is rarely made specific and is highly questionable. Future costs should first be adjusted for inflation using an estimate of the rate of inflation that is specific to the cost sector. That is, medical care costs should be adjusted for inflation in the medical care cost sector. The inflated costs should only then be discounted (Weinstein and Fineberg 1980).

There is a second issue concerning inflation. Available data on charge or payment frequently come from data sources developed in different years. For example, information on the cost of surgery may be available only for the year 1988, whereas information on the cost of medical treatment is available for 1991. Cost information should always be adjusted for inflation so that all costs are measured in current dollars at the start of the analysis. The rate of inflation used in this adjustment should approximate the measured inflation rate in the relevant cost sector. That is, hospital costs should be inflated by the rate of inflation of hospital costs and pharmaceutical costs should be inflated by the rate of inflation of pharmaceuticals.

12.6 DISCOUNTING BENEFITS

When discounting costs, it is recommended that health benefits (i.e., years of life saved) also be discounted at the same rate (Drummond, Stoddart, Torrance 1987; Keeler and Cretin 1983). When benefits are discounted, years of life saved in the future are valued less than years of life saved earlier. The rationale for discounting costs is *not* a desire to demean the value of future life (Weinstein and Fineberg 1980). Rather, because lives, years of life saved, and other utilities are being valued relative to dollars, and dollars are discounted to their present value, it is necessary to discount benefits at the same rate (Weinstein and Fineberg 1980).

In addition, Keeler and Cretin (1983) show how the failure to discount benefits when costs are discounted leads to logical inconsistencies. Specifically when costs are discounted and benefits are not discounted, for any program begun now, there is always a delayed program which should be funded first, and no program with a finite starting date can be selected. It is thus impossible to make a decision to begin a program based on its being cost effective.

The value of the discount rate for benefits may greatly affect conclusions based on a cost-effectiveness analysis.

EXAMPLE: Table 12-7 shows the cost per added year of life for a childhood cholesterol screening program compared with no screening using several combinations of the discount rate for costs and benefits. The cost per added year of life is least when benefits are not discounted. A conclusion about the "cost effectiveness" of the program is dependent on the choice of the discount rate for benefits.

Table 12-7 Cost per added year of life for a childhood cholesterol screening program under two discounting assumptions

Discount Rate (%)		Cost per Year of Life
Costs	Benefits	
5	0	$1,000
5	5	9,400

Source: Keeler and Cretin (1983).

There are several different methods for discounting benefits, and the different methods for discounting benefits give different results (Johannesson 1992).

EXAMPLE: Table 12-8 compares different methods of discounting benefits for a hypothetical therapy that would reduce overall mortality by 50% every year for 8 years (Johannesson 1992). It compares the effect of the therapy on life expectancy for 65-year-old and 45-year-old men. If the cost of the therapy is $10,000 per year for 8 years, the cost per year of life gained in a 65-year-old man ranges from $54,058 to $74,562 depending on the method used to discount benefits. For the 45-year-old man, the range of cost per year of life gained is even wider, depending on the method used to discount benefits.

There is no consensus on the correct method for discounting benefits. With or without discounting of benefits, the cost-effectiveness analysis does not answer the

Table 12-8 Comparison of the discounted gain in life expectancy and cost per year of life gained as estimated using three methods for discounting, for a hypothetical intervention that reduces overall mortality by 50% for eight years

Method for Discounting Benefits	Gain in Life Expectancy (years)		Cost per Year of Life Gained[a]	
	45-year-old	65-year-old	45-year-old	65-year-old
Method 1[b]	0.22	0.87	$306,309	$74,562
Method 2[c]	0.23	0.94	292,991	69,010
Method 3[d]	0.41	1.20	164,361	54,058

[a]Both costs and benefits are discounted at 5%.

[b]Life expectancy is calculated as the sum of cumulative survival probabilities. These probabilities are discounted before they are added together to estimate life expectancy with and without the intervention. The gain in life expectancy with and without the intervention is then calculated as the difference between the discounted life expectancies.

[c]The starting point for discounting is the year that the gain in life expectancy occurs. The gain in life expectancy in one year is multiplied by the discounted remaining life expectancy and then multiplied by the probability of surviving to that year.

[d]The method is the same as Method 2 except that remaining life expectancy is not discounted.

Source: Johannesson (1992).

question of whether it is "worth it" to spend $54,058 to gain a year of life in a 65-year-old man.

The large differences in the estimates of the cost per year of life gained that can arise because benefits are discounted are particularly troublesome because the discount rate for benefits is chosen rather arbitrarily. When a decision is made to discount benefits, a sensitivity analysis of the discount rate for benefits should always be done.

12.7 TIME HORIZON

The time horizon of an analysis is a time in the future beyond which all costs and benefits are ignored (Sugden and Williams 1990). It is sometimes useful to define a short time horizon for a cost-effectiveness analysis to mitigate problems that grow out of discounting, inflation, and the existence of differences in the rate of inflation and the general rate of inflation. Defining a short time horizon also obviates the difficulties that arise from lack of knowledge about what benefits will be far in the future.

EXAMPLE: In large measure to avoid the problem of discounting, we chose to ignore costs of caring for premature infants after their discharge from the hospital in our analysis that attempted to estimate the break-even cost of smoking cessation programs in pregnancy (Shipp et al. 1992).

13

Sensitivity Analysis

Sensitivity analysis is an essential element of decision analysis and cost-effectiveness analysis. The principles of sensitivity analysis are also applicable to meta-analysis. This chapter shows how to do sensitivity analysis for studies using each of the three methods.

Section 13.1 describes the overall purpose of sensitivity analysis. Section 13.2 describes one-way sensitivity analysis as applied to decision analysis and cost-effectiveness analysis. Section 13.3 describes how to do and how to interpret two-way and three-way sensitivity analysis. It discusses *n*-way sensitivity analysis. Section 13.4 describes the application of the principles of sensitivity analysis to meta-analysis.

13.1 GOALS OF SENSITIVITY ANALYSIS

Sensitivity analysis evaluates the stability of the conclusions of an analysis to assumptions made in the analysis. When a conclusion is shown to be invariate to the assumptions, confidence in the validity of the conclusions of the analysis is enhanced.

Sensitivity analysis also helps identify the most critical assumptions of the analysis. This knowledge can be used to formulate priorities for future research aimed at resolving the problem posed in the analysis.

Table 13-1   One-way sensitivity analysis varying estimates of the operative mortality rate for prophylactic cholecystectomy in decision analysis comparing prophylactic cholecystectomy and expectant management for silent gallstones

Probability	Assumed Value	Gain (Loss) in Life Expectancy for a 30-year-old Man[a]
Operative mortality	0.107 (baseline)	(4)
	0.428	(17)
	0.000	7

[a]For cholecystectomy.
Source: Ransohoff et al. (1983).

13.2   ONE-WAY SENSITIVITY ANALYSIS IN DECISION ANALYSIS AND COST-EFFECTIVENESS ANALYSIS

13.2.1   Overview

An implicit assumption of decision analysis is that the values of the probabilities and of the utility measure are the correct values for these variables. In cost-effectiveness analysis, there is a similar implicit assumption that the discount rate and the costs are correct. In one-way sensitivity analysis, the assumed values of each variable in the analysis are varied, one at a time, while the values of the other variables in the analysis remain fixed. One-way sensitivity analysis of a cost-effectiveness analysis should also vary the discount rate for costs and benefits while keeping the values of the other variables in the analysis fixed.

> *EXAMPLE:*   Ransohoff et al. (1983) did a decision analysis comparing prophylactic cholecystectomy with expectant management (waiting for symptoms) in asymptomatic patients diagnosed with gallstones. Using the baseline probabilities, the analysis showed that a 30-year-old man would loose 4 days of life by choosing prophylactic cholecystectomy. Table 13-1 presents the results of a one-way sensitivity analysis in which the value of the operative mortality rate was varied while the values of the other variables were held constant. The analysis shows prophylactic cholecystectomy would increase life expectancy by 7 days if operative mortality was zero and would decrease it by 17 days if operative mortality was four times the baseline estimate.

13.2.2   Interpreting the Results of One-Way Sensitivity Analysis

When the assumed value of a variable affects the conclusion of the analysis, the analysis is said to be "sensitive" to that variable. When the conclusion does not change when the sensitivity analysis includes the values of the variables that are within a reasonable range, the analysis is said to "insensitive" to that variable.

If an analysis is sensitive to the assumed value of a variable, the likelihood that the extreme value is the true value can be assessed qualitatively. The absolute amount of benefit of one strategy over the other under the extreme assumptions can be weighed. Further research to refine the estimate may be a priority for future studies.

   EXAMPLE: In the one-way sensitivity analysis for the problem of prophylactic cholecystectomy compared with expectant management of silent gallstones that was described in Section 13.1.2, prophylactic cholecystectomy was estimated to decrease life expectancy in a 30-year-old man by 4 days if the operative mortality was 0.107, the baseline estimate, but it would increase life expectancy by 7 days if mortality was assumed to be 0. The analysis is *sensitive* to the assumptions about the operative mortality rate. However, since it is unlikely that mortality could be reduced to zero and since the gain in life expectancy is small even with a zero operative mortality rate, the strategy of watchful waiting is favored.

   When the analysis is sensitive to the utility measure, the choice of the utility measure must be justified carefully. When the conclusion of a cost-effectiveness analysis is sensitive to the discount rate, the choice of the discount rate must be explained. The implications of the choice of the discount rate for interpretation of the analysis can be explained. It may be possible to redefine the time horizon for the analysis to avoid discounting.

13.2.3 Threshold Analysis

Threshold analysis is an extension of one-way sensitivity analysis. In threshold analysis, the value of one variable is varied until the alternative decision strategies are found to have equal outcomes, and there is no benefit of one alternative over the other in terms of estimated outcome. The threshold point is also called the "break-even" point. At the break-even point, the decision is a "toss-up" (Kassirer and Pauker 1981). That is, neither of the alternative decision options being compared is clearly favored over the other.

   Threshold analysis is especially useful when consideration is being given to the use of an intervention in groups that can be defined a priori based on the values of the variable that is the subject of the threshold analysis.

   EXAMPLE: In the decision analysis of isoniazid prophylaxis in HIV-infected intravenous drug users by Jordan et al. (1991) that was discussed in Chapter 1 and Chapter 9, the overall conclusion was that prophylaxis was beneficial in all groups except black women with negative tuberculin skin tests. One of the key probabilities is the life expectancy in the absence of tuberculosis. An analysis was done to determine the threshold for the benefit of isoniazid prophylaxis. Table 13-2 shows the threshold values. For each of the groups defined in the table, the threshold value is the value of life expectancy at which the strategy of giving prophylaxis would be equal to no prophylaxis in terms of additional life expectancy, considering prevention of tuberculosis. The analysis shows that in all the groups except black women, who had no benefit under baseline assumptions, even persons with a low remaining life expectancy would benefit from tuberculosis prophylaxis with isoniazid.

Table 13-2 Threshold analysis for isoniazid in HIV-infected intravenous drug users: Values of years of remaining life expectancy that would make the decision to use isoniazid a "toss-up"

	Tuberculin Test Result	
Patient Description	Positive	Negative
Black men	3.31	5.29
Black women	3.44	No benefit
White men	3.28	4.50
White women	3.27	4.32

Source: Jordan et al. (1991).

13.3 TWO-WAY, THREE-WAY, AND *n*-WAY SENSITIVITY ANALYSIS

13.3.1 Two-Way Sensitivity Analysis

In two-way sensitivity analysis, the expected outcome is determined for every possible combination of reasonable estimates of two variables, while the values of all other variables in the analysis are held constant at baseline. In two-way sensitivity analysis, it is usual to identify the pairs of values that equalize the expected outcome or expected utility of the alternatives and to present the results of the analysis graphically. It is difficult to interpret the results of a two-way sensitivity analysis without the aid of graphs.

EXAMPLE: In the decision analysis comparing the utility of amniocentesis and chorionic villus sampling discussed in Chapter 10, (Heckerling and Verp 1991), a two-way sensitivity analysis varying the estimates of spontaneous abortion following amniocentesis and following chorionic villus sampling was done. Figure 13-1 graphically presents the results of a two-way sensitivity analysis. In the graph, the solid line appears at combinations of the rates of spontaneous abortion that equalize the expected utility of amniocentesis compared with chorionic villus sampling. The combinations of spontaneous abortion rates that would lead to a preference for chorionic villus sampling are shown with dark shading. The combinations that would lead to a preference for amniocentesis are shown with light shading. The baseline estimate shown with a dot.

The two-way sensitivity analysis shows that, with the rate of spontaneous abortion after amniocentesis fixed at its baseline rate of 2.8%, amniocentesis is the preferred strategy if the rate of spontaneous abortion after chorionic villus sampling is greater than 2.68%. Amniocentesis is the preferred strategy at all combinations where the spontaneous abortion rate following chorionic villus sampling is greater than the spontaneous abortion rate following amniocentesis. Since chorionic villus sampling is probably inherently more likely to cause spontaneous abortion than amniocentesis, the two-way sensitivity analysis strengthens confidence in the overall conclusion of the analysis favoring amniocentesis over chorionic villus sampling.

13.3.2 Three-Way Sensitivity Analysis

In three-way sensitivity analysis, the expected outcome is determined for combination of reasonable estimates of three variables, while the values of all other variables in the analysis are held constant at baseline. Like two-way sensitivity analysis, it is usual to present the results of three-way sensitivity analysis graphically, and interpretation of the results in the absence of graphical aids is difficult.

EXAMPLE: Heckerling and Verp (1991) did a three-way sensitivity analysis. In it, they varied the probability of spontaneous abortion after cho-

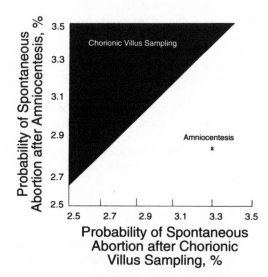

Figure 13-1 Two-way sensitivity analysis from the decision analysis by Heckerling and Verp (1991) comparing amniocentesis and chorionic villus sampling. The probabilities that were subjected to two-way sensitivity analysis are the probability of spontaneous abortion after amniocentesis and the probability of spontaneous abortion after chorionic villus sampling. Other probabilities were held at their baseline values.

The region to the right of the line represents combinations of probabilities of spontaneous abortion for which amniocentesis would be preferred; the region to the left represents combinations for which chorionic villus sampling would be preferred. The baseline combination of values is shown with an x. (Reproduced with permission from Heckerling and Verp, *Journal of Clinical Epidemiology*, 1991;44:663.)

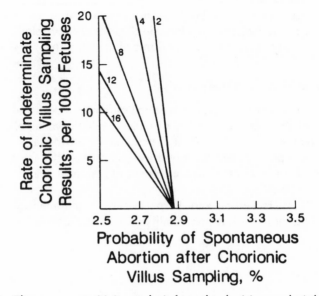

Figure 13-2   Three-way sensitivity analysis from the decision analysis by Heckerling and Verp (1991) comparing amniocentesis and chorionic villus sampling. The probabilities that are subjected to three-way sensitivity analysis are the probability of spontaneous abortion after chorionic villus sampling, the rate of indeterminate chorionic villus sampling results, and the probability of an abnormal amniocentesis after an indeterminate chorionic villus sampling. The other probabilities are held at their baseline values.

For any line, the region to the right represents combinations of the probability of spontaneous abortion after chorionic villus sampling and the rate of indeterminate chorionic villus sampling results for which amniocentesis is preferred; the region to the left represents combinations for which chorionic villus sampling is preferred. (Reproduced with permission from Heckerling and Verp, *Journal of Clinical Epidemiology*, 1991;44:663.)

rionic villus sampling, the rate of indeterminate chorionic villus sampling, and the probability of an abnormal amniocentesis after an indeterminate chorionic villus sampling, while holding the values of other variables in the analysis constant at their baseline levels. Figure 13-2 depicts the results graphically. Five lines are shown for five values of the probability of an abnormal amniocentesis following indeterminate chorionic villus sampling. For each of these values, the line shows the combination of values of the rate of indeterminate chorionic villus sampling and probability of spontaneous abortion after chorionic villus sampling that would yield the same expected utility for amniocentesis and chorionic villus sampling. Choosing one line, the combinations of values to the right of the line represent values for which amniocentesis is preferred and the combinations to the left represent values for which chorionic villus sampling is preferred. Figure 13-2 is analogous to Figure 13-1. In fact, the results could have been presented as five figures like Figure 13-1, one for each value of the probability of an abnormal amniocentesis following indeterminate chorionic villus sampling.

The three-way sensitivity analysis shows that amniocentesis is the preferred strategy except if the rate-indeterminate chorionic villus sampling and the spontaneous abortion rate after chorionic villus sampling are implausibly low and the rate of abnormal amniocentesis implausibly high. The sensitivity analysis further strengthens confidence in the conclusion of the decision analysis favoring amniocentesis over chorionic villus sampling.

13.3.3 n-Way Sensitivity Analysis

In n-way sensitivity analysis, the expected outcome is determined for every possible combination of every reasonable value of every variable. n-Way sensitivity analysis is analogous to n-way regression. It is difficult to do and difficult to interpret. n-Way sensitivity analysis will not be described further in this book.

13.3.4 Choice of Variables for Sensitivity Analysis

It is usual to do one-way sensitivity analysis for each variable in the analysis. The highest and the lowest values within the reasonable range of values are first substituted for the baseline estimate in the decision tree. If substitution of the highest or the lowest value changes the conclusions, more values within the range are substituted to determine the range of values for which the conclusions apply.

It is especially important to do one-way sensitivity analysis for the discount rate in cost-effectiveness analysis, because there is so much controversy about the choice of a discount rate. If the conclusion of a cost-effectiveness analysis is shown to be independent of the choice of the discount rate within a reasonable range of estimates of the discount rate, then arguments about the appropriateness of discounting and about the proper discount rate are moot.

EXAMPLE: Chapter 12 described a cost-effectiveness analysis of bone marrow transplantation in acute nonlymphocytic leukemia. Table 13-3 compares both the cost of bone marrow transplantation and the cost of chemotherapy with doing nothing for several estimates of the discount rate. For

Table 13-3 For several estimates of the discount rate, cost per year of life saved for bone marrow transplantation compared with doing nothing and for chemotherapy compared with doing nothing in patients with acute nonlymphocytic leukemia

Discount Rate (%)	Cost per Year of Life Gained	
	Chemotherapy	Bone Marrow Transplant
0	$10,300	$ 6,900
5	22,900	16,600
10	35,500	27,900

Source: Welch and Larson (1989).

all of the values from 0 to 10%, bone marrow transplantation costs less than chemotherapy. Thus, bone marrow transplantation is cost effective relative to chemotherapy by the definition given in Chapter 12. This conclusion does not depend on assumptions about the discount rate.

In an analysis that uses measures of utility derived by any of the methods described in Chapter 11, it is important to do a sensitivity analysis varying the utility. If the analysis is insensitive to assumptions about the utility within reasonable estimates of the utility measure, criticisms of the utility measure itself carry less weight.

In an analysis with many probabilities, there are numerous combinations of two variables and three variables, and the computational burden of doing all possible two-way and three-way sensitivity analyses is large. For this reason, it is not usually feasible to do two-way sensitivity analysis for all combinations of two variables. The choice of variables for two-way and three-way sensitivity analysis requires considerable judgment, and there are no hard and fast rules. The variables that seem the most controversial may be chosen for two-way sensitivity analysis, since the believability of the conclusions of the analysis may hinge on assumptions about these variables.

EXAMPLE: The rates of spontaneous abortion following chorionic villus sampling and amniocentesis are controversial. However, there is general agreement that the rate of spontaneous abortion is higher for chorionic villus sampling than for amniocentesis. The two-way sensitivity analysis showed that the rate of spontaneous abortion following amniocentesis would have to be *higher* than the rate following chorionic villus sampling to alter the conclusion favoring amniocentesis. The confidence in the conclusion is enhanced through this two-way sensitivity analysis.

13.4 APPLICATION OF THE PRINCIPLES OF SENSITIVITY ANALYSIS TO META-ANALYSIS

There are several ways in which the principles of sensitivity analysis are applicable in meta-analysis. First, the stability of the conclusions of the meta-analysis to assumptions about the studies that should be included and excluded from the meta-analysis can be carried out.

EXAMPLE: Chapter 6 described a meta-analysis of studies of coronary heart disease and estrogen use done by Stampfer and Colditz (1991). One of the studies that was eligible for this meta-analysis, the study by Wilson, Garrison, and Castelli (1985), was controversial, because the authors adjusted their estimate of the risk of coronary heart disease in estrogen users for differences in HDL cholesterol between estrogen users and nonusers. Because changes in HDL cholesterol may mediate the beneficial effect of estrogen on coronary heart disease, adjustment for HDL cholesterol constitutes over-

Table 13-4 Summary estimate of the relative risk of coronary heart disease in estrogen users based on cohort studies with internal controls including and excluding one controversial study

Analysis	Summary Estimate of Relative Risk (95% confidence interval)
Including Wilson et al. (1985)	0.58 (0.48–0.69)
Excluding Wilson et al. (1985)	0.56 (0.47–0.67)

adjustment. A second analysis on the topic based on data from the Framingham Study was published in 1987 and came to a different conclusion (Eaker and Castelli 1987). The results of the second analysis were poorly described, and the information from the second analysis was not presented in a form suitable for inclusion in a meta-analysis. Table 13-4 shows that the summary estimate of the relative risk of coronary heart disease in estrogen users is not affected at all by the decision to include or exclude the data from the analysis of Wilson et al. (1985). Thus, attempts to discredit the Wilson et al. analysis as a way of justifying excluding the results are a waste of space and energy.

An analysis examining the stability of conclusions to assumptions about the effect size in each study may be done.

EXAMPLE: Section 6.9 described the study by Bush et al. (1987), which was eligible for the meta-analysis of Stampfer and Colditz (1991). It reported the results of many different analyses with adjustments for many different sets of potential confounders. Table 13-5 summarizes estimates of the relative risk of coronary heart disease in estrogen users based on cohort studies in which the different estimates of the relative risk of disease in estrogen

Table 13-5 Summary estimates of the relative risk of coronary heart disease in estrogen users based on cohort studies with internal controls varying the estimate of relative risk from the study of Bush et al. (1987) while holding other estimates constant

Estimated Relative Risk from Study by Bush et al. (1987) (95% confidence interval)	Summary Estimate of Relative Risk[a] (95% confidence interval)
0.37 (0.16–0.88)[b]	0.58 (0.48–0.69)
0.34 (0.12–0.81)	0.58 (0.49–0.70)
0.44 (0.19–1.03)	0.58 (0.49–0.70)
0.47 (0.20–1.12)	0.58 (0.49–0.70)
0.21 (0.00–0.51)	0.59 (0.49–0.71)
0.48 (0.00–1.00)	0.59 (0.49–0.71)
0.42 (0.13–1.00)	0.59 (0.49–0.70)

[a]Based on all cohort studies with internal controls.

[b]Estimate that was used in analysis reported by Stampfer et al. (1991).

users from the study by Bush and colleagues are varied while other estimates of relative risk are held constant. The analysis shows that the choice of the risk estimate from the Bush et al. study does not materially affect the conclusion of the meta-analysis. This sensitivity analysis thus strengthens confidence in the validity of the overall conclusion that estrogen replacement therapy prevents coronary heart disease.

Last, the assumptions made in the analytic model used in the analysis can be assessed.

EXAMPLE: The meta-analysis of environmental tobacco smoke and lung cancer discussed in Chapter 1 (United States Environmental Protection Agency 1990) used the Mantel-Haenszel method, a method based on the assumption of fixed effect. The estimated relative risk of lung cancer in women exposed to environmental tobacco smoke was 1.42 (95% C.I. 1.24–1.63) when based on this method. A test of heterogeneity yielded a probability value greater than 0.05, and this result suggests that a summary estimate is valid. Analysis of data from these studies using the random-effects model as described in Chapter 7 yields an estimated relative risk of 1.35 (95% C.I. 1.18–1.55). The estimates do not differ by much, and a conclusion about the statistical significance of the elevation in the relative risk based on inspection of the confidence interval is the same for the analysis based on the fixed-effects model and the random-effects model. The similarity of estimates based on the two models suggests that model choice is not an issue.

14

Reporting Results

The published report of the results of a meta-analysis, decision analysis, or cost-effectiveness analysis is usually the only information about the study that is available to readers. Most readers do not have the technical expertise to identify all of the assumptions of the study. These methods are mostly unfamiliar and they seem complex. Skepticism about the conclusions of the analysis may be high because of the unfamiliarity of the methods and the fact that they require complex quantitative analysis. For all of these reasons, the description of the study methods and procedures must be comprehensive, and the presentation of the study findings must be especially clear. Devices that make the results easy to understand, such as graphs and charts, are almost always a necessary part of the description of these studies and the presentation of their results.

Sections 14.1 to 14.3 describe the specific information that should be included in the published reports of each of three types of studies. Sections 14.4 and 14.5 describe some of the graphical techniques that can be used to simplify the presentation of results of the studies that use these methods.

14.1 META-ANALYSIS

14.1.1 Identification of Eligible Studies

The details of the procedures that were used to identify studies eligible for the meta-analysis should be given. The report should describe the exact terms that were used to search MEDLINE and other computer-stored databases. If individual authors were asked to furnish unpublished studies or to clarify results, the methods of contact should be described, and the response rate should be given.

The criteria for eligibility of studies identified by the information retrieval process should be specified, along with the rationale for each criterion, if this is not obvious. All of the studies with relevant material that were identified should be cited in the report, and minimal information on both the included and the excluded studies (author, year of publication, number of subjects) should be presented in a table. For each excluded study, the reason for exclusion should be presented alongside the citation or in a footnote. Descriptive information on the characteristics of subjects in the studies included in the meta-analysis (e.g., age, sex, race) should be provided.

14.1.2   Quality Rating

If the studies were rated on quality, the methods for obtaining the ratings should be described. The items that make up the scale should be described. The results of assessments of the reliability and validity of the quality rating scale should be presented.

14.1.3   Data Abstraction

The report should describe the procedures for abstracting data from the study reports. If abstractors were blinded to various aspects of the publication while abstracting other portions, this should be stated. If abstractors were not blinded, this fact should also be stated.

If a quality score was assigned, the methods for abstracting data to assign a quality score of studies should also be described. A statement of whether the assessment of quality was done blind to the results or other aspects of the study, such as the author and journal of publication, should be a part of this description.

Steps taken to ensure reliability of data abstraction, such as training of the abstractors or reabstraction of a sample of records, should be described. The results of any formal evaluations of the reliability of data abstraction should be presented, even if reliability was not very good.

The rules for choosing among estimates of effect when there is more than one should be stated. The procedures for handling missing data should be described.

14.1.4   Statistical Analysis

The report should explain the reason for the choice of the effect measure used in the analysis. It should state whether the analysis was based on a fixed-effects or random-effects model. The rationale for the choice of model should be provided. The method for arriving at the summary estimate of effect and its confidence limit should be described either directly or by appropriate citation to the published literature.

The test for assessing statistical lack of homogeneity should be described along with the results of the test. When there is statistical evidence of lack of homogeneity, the effect of lack of homogeneity on conclusions and on the choice of model should be discussed.

If quality of the studies was rated, the method for taking it into account in the analysis should be described.

14.1.5 Discussion

The possibility of publication bias should be addressed. If there is any evidence for and against publication bias as an explanation for the results, the evidence should be presented.

14.2 DECISION ANALYSIS

14.2.1 Decision Tree

The decision tree should be presented graphically. All of the symbols used in the tree should be explained in a legend. The tree should be constructed following the conventions that are described in Chapter 4.

14.2.2 Probability Estimates

The source of all the probability estimates should be cited. When data from several studies have been aggregated to obtain a probability estimate, the method for aggregating the data should be described. When there is more than one estimate for a given probability and one estimate is chosen for the decision analysis, the reasons for choosing this estimate should be given. A table that shows the range of estimates for each probability is useful.

14.2.3 Measure of Utility

If the analysis used a measure of utility, the methods for measuring the utility should be described. This description should include the number of subjects, the source of the subjects, and the method for collecting the data. Information on the reliability and validity of the utility measure should be provided, if it is available.

14.2.4 Sensitivity Analysis

The variables that were subjected to sensitivity analysis should be identified and the reasons for doing a sensitivity analysis on these variables should be given. The range used for the sensitivity analysis of each variable should be justified. The results of the sensitivity analysis should be presented so that readers can understand them.

14.2.5 Discussion

The discussion section should describe with honesty the limitations of the analysis. Critical assumptions and uncertainties should be identified. The importance of the uncertainties and assumptions to the overall conclusion of the analysis should be considered candidly.

If other analyses of the same topic have been done, the differences and similarities between the analyses should be described. An attempt should be made to explain reasons for differences.

14.3   COST-EFFECTIVENESS ANALYSIS

14.3.1   Perspective

The perspective of the analysis should be stated. The reason for the choice of perspective should be given.

14.3.2   Time Horizon

The time horizon for the study should be stated. The rationale for the choice of the time horizon should be given.

14.3.3   Cost Data

The source of cost estimates should be given. If data on cost were collected in a special study, the methods of the study should be described. If charge data are used as a proxy for cost, the rationale for this decision should be given.

14.3.4   Discount Rate

The discount rates used for costs and for effects should be given. The reason for the choice of the discount rate should be provided. The results of a sensitivity analysis varying the discount rate should be presented.

14.4   GRAPHICAL PRESENTATION OF THE RESULTS OF META-ANALYSIS

14.4.1   Overview

The function of a graph is to convey an immediate impression of the relationships among the numbers (Gladen and Rogan 1983). The format of a graph can convey markedly different impressions of the relationships, some erroneous. Attention to the format of graphs used to present the results of meta-analysis is important.

14.4.2   Ratio Measures

Figure 14-1 is an example of the graphic used most frequently to present the results of a meta-analysis of studies where the effect is measured on a ratio scale. Data from the five hospital-based case-control studies of coronary heart disease in estrogen users included in the meta-analysis of Stampfer et al. (1991) that was

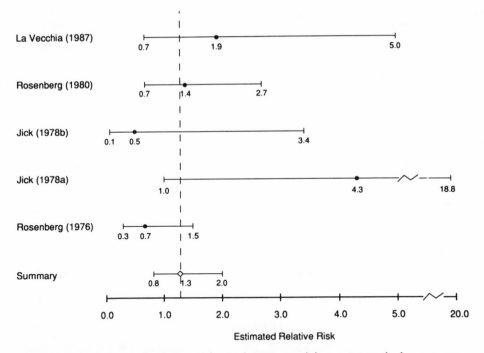

Figure 14-1 Estimated relative risks and 95% confidence intervals for coronary heart disease in estrogen users from hospital-based case-control studies. The relative risk estimates and the 95% confidence intervals are plotted on a linear scale. Data and individual references are from Stampfer et al. (1991).

discussed in Chapter 6 are presented. The meta-analysis of studies of breast cancer treatments that was discussed in Chapter 1 was presented in an identical type of figure. Each study in the meta-analysis is represented by one horizontal line, whose length is proportional to the length of the 95% confidence interval; the point estimate of effect in the study is indicated with a large dot. The figure also includes a line for the summary estimate of relative risk and its 95% confidence interval. A dashed line is drawn vertically through the summary estimate of relative risk such that it crosses each of the lines representing the individual studies.

Figure 14-1 shows the estimated relative risks and 95% confidence intervals plotted on a linear scale. There are several reasons why plotting ratio measures on a linear scale is unsatisfactory (Gladen and Rogan 1983; Galbraith 1988; Hebert and Miller 1989). First, when plotted on a linear scale, values of the ratio measure and its reciprocal, which are equivalent, are not equidistant from 1.0. Second, a unit change in the ratio measure does not have the same interpretation at all points of the scale. For example, a change in the estimated relative risk ratio from 2.0 to 3.0 is a 50% increase, whereas a change from 4.0 to 5.0 is only a 25% increase. Third, when plotted on a linear scale, studies with estimated relative risks between 0.0 and 1.0 appear to be less important than studies with estimated relative risk above 1.0 because they take up less visual space on the page. A number of authors have suggested ways to plot ratio measures that remedy these problems

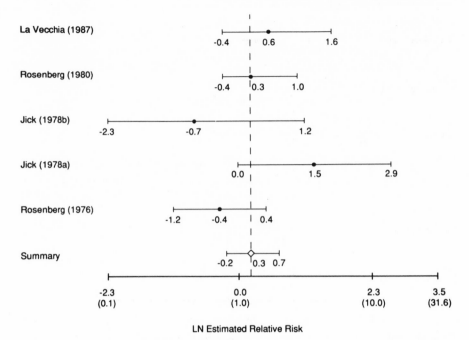

Figure 14-2   Data from Figure 14-1. The natural logarithm of the relative risk estimates and the natural logarithms of 95% confidence intervals are plotted on a linear
scale.

(Gladen and Rogan 1983; Greenland 1987; Galbraith 1988; Hebert and Miller
1989; Morgenstern and Greenland 1990).

First, the ratio measure and its confidence interval can be plotted on a logarithmic scale (Gladen and Rogan 1983; Galbraith 1988).

EXAMPLE: In Figure 14-2 the logarithms of data from Figure 14-1 are
plotted on a linear scale. The data could have been plotted directly on a
logarithmic scale. Plots on a logarithmic scale and plots of the logarithm are
superior to plots on a linear scale because the studies that yield estimates of
the relative risk of coronary heart disease in estrogen users less than 1.0 and
studies that yield estimates greater than 1.0 are equidistant from 1.0, and
changes in the estimated relative risk of equal size are represented equally.

Plotting the ratio measure on a reciprocal scale also yields a graph where estimates less than 1.0 and greater than 1.0 are equidistant from 1.0. Morgenstern
and Greenland (1990) describe the special advantage of the reciprocal plot as a
way to convey proportional impact. Unlike the logarithmic plot, equal ratios
between pairs of estimates do not translate into equal distances on a reciprocal
plot.

EXAMPLE: Figure 14-3 is a reciprocal plot of the data shown in Figures
14-1 and 14-2.

Plotting the 95% confidence intervals on either a linear or logarithmic scale tends to draw visual attention to the studies that are the least precise and have the lowest weight in the analysis, since less precise estimates have longer confidence intervals and take up more visual space in the plot (Morgenstern and Greenland 1990). To give greater visual prominence to studies with more weight in the analysis, a box that has an area equal to the statistical weight of the study can be drawn around the point estimate of relative risk. In such plots, the eye is drawn to the study with the greatest weight in the analysis.

EXAMPLE: Figure 14-4 shows the data from the study of Stampfer et al. (1991) plotted on a linear scale with the size of box around the point estimate of relative risk equal in area to the statistical weight of the study in the analysis.

Galbraith (1988) describes the radial plot. It has the same advantages as plots on a logarithmic scale. Radial plots are difficult to construct and fairly hard to read, and they are seldom used in practice.

Weighted histograms (Greenland 1987) can be useful for some problems. The odd-man-out procedure (Walker, Martin-Moreno, Artalejo 1988) is a graphical method for estimating the confidence interval for a meta-analysis. Neither of these methods has been used much in practice, and they will not be discussed further.

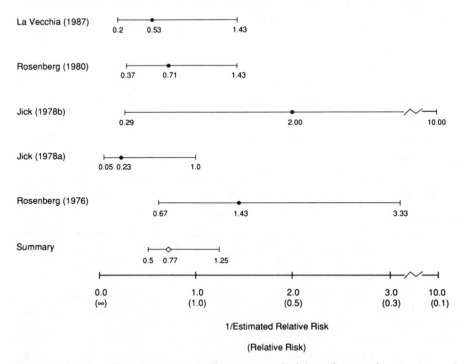

Figure 14-3 Data from Figure 14-1. The reciprocal of the relative risk estimates and the reciprocals of the 95% confidence intervals are plotted on a linear scale.

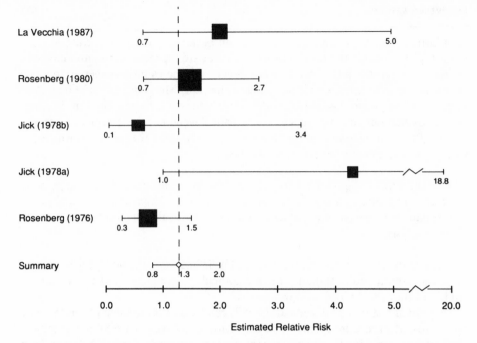

Figure 14-4 Data from Figure 14-1. The relative risk estimates and the 95% confidence intervals are plotted on a linear scale, but the point estimate of relative risk is depicted using a box whose size is proportional to the weight of the study in the meta-analysis. Studies yielding more precise estimates of relative risk are depicted with boxes that are larger than studies yielding less precise estimates of relative risk.

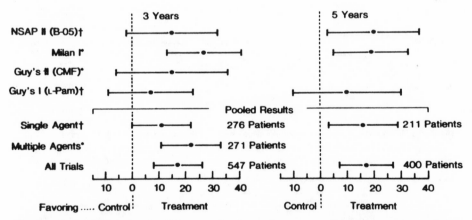

Figure 14-5 Difference in the rate of relapse-free survival at three years and five years among premenopausal women in various trials of chemotherapy for breast cancer. Each line represents one study. The point estimate of the rate difference is depicted with a filled circle; the length of the line is the length of the 95% confidence interval for the rate difference. References to individual studies are as cited in Himel et al. (1986). (Reproduced with permission from Himel et al., *Journal of the American Medical Association,* 1986;256,1157.)

14.4.3 Difference Measures

When effect is measured as a difference between two dichotomous outcomes—death or nondeath, survival or nonsurvival—the graph used to present the results of the analysis is a difference scale. Plotting rate differences on a linear scale does not compress studies with negative differences into a small visual space.

> *EXAMPLE:* Figure 14-5 shows the rate differences in relapse-free survival of postmenopausal breast cancer patients treated or not treated with adjuvant chemotherapy from an analysis by Himel et al. (1986).

As with ratio measures, this visual presentation of the difference measure tends to draw the eye to studies with the widest confidence intervals, which are the least

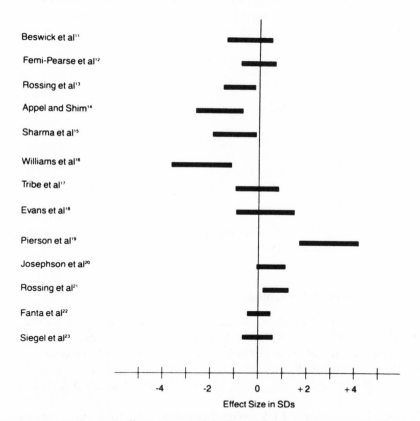

Figure 14-6 Estimated effect of aminophylline treatment compared with control where effect size is expressed in standardized units. Each bar represents one study. Positive effect sizes indicate aminophylline treatment was more effective than control; negative effect sizes indicate the control was more effective than aminophylline. The length of each bar is the length of the 95% confidence interval of the standardized measure for that study. References to individual studies are as cited in Littenberg (1988). (Reproduced with permission from Littenberg, *Journal of the American Medical Association*, 1988;259,1680.)

Table 14-1   Estimates of the relative risk of coronary heart disease and physical inactivity by quality of study

Studies of Poor Quality ($N = 19$)	Studies of Good Quality ($N = 13$)	Studies of Best Quality ($N = 9$)
0.5 Low extreme		
0.5	1.2 Low extreme	1.2 Low extreme
0.9	1.5	1.4
1.1	1.6	1.6
1.1	1.6[a]	1.6[a]
1.1[a]	1.8	1.9 Median
1.1	1.9	2.4
1.1	2.0 Median	2.5[a]
1.5	2.0	2.6
1.5 Median	2.0	3.1 High extreme
1.5	2.0[a]	
1.5	2.3	
1.7	2.3	
1.8	2.5	
1.9[a]	2.8 High extreme	
2.0		
2.2		
2.3		
2.4		
2.5 High extreme		

[a] Values that define 50% of estimates.

Source: Williamson, Parker, Kendrick (1989).

precise. Presenting a box that has an area equal to the statistical weight of the study would mitigate this problem.

14.4.4   Continuous Variables

When a treatment or intervention affects a continuous measure, such as blood pressure or weight, the plot can show effect size as the measured difference between the treated and the control group just as described for differences in a dichotomous measure. If the measures of effect used in different studies are measured on different scales, they can be plotted in units of standard deviation.

EXAMPLE: Figure 14-6 depicts the results of studies of the effect of aminophylline treatment in acute asthma that were included in the meta-analysis of this topic by Littenberg (1988) that was discussed in Chapter 8. Since studies used different measures of pulmonary function, effects were standardized in units of the standard deviation. These have been plotted on a scale whose *x*-axis is units of standard deviation. The length of the bars is equal to the 95% confidence interval for the measure of standard deviation.

14.4.5 Box Plots

Williamson, Parker, and Kendrick (1989) have pointed out the usefulness of the box plot in the presentation of the results of meta-analysis. The box plot can be used for ratio measures, difference measures, and continuous measures.

The box plot, which is also called a box-and-whiskers plot, plots the median, the approximate quartiles, and the highest and lowest data points as a way of conveying visually the level, the spread, and the amount of symmetry of the data points.

EXAMPLE: Table 14-1 reproduces the data from a meta-analysis of studies of physical activity and coronary heart disease that Williamson, Parker, and Kendrick (1989) to illustrate the presentation of the results of a meta-analysis through a box plot. Figure 14-7 shows box plots of studies in

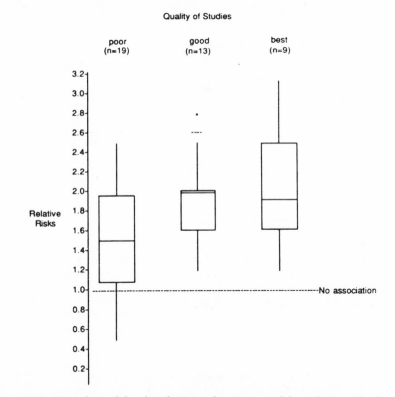

Figure 14-7 Box plots of the distribution of estimates of the relative risk of coronary heart disease in physically inactive compared with physically active men in studies rated as poor, good, and best in quality. The horizontal line inside the box is the median value of the relative risk. The upper and lower ends of the box are the "hinges," the approximate upper and lower quartiles of the distribution of relative risks. The vertical lines from the ends of the boxes connect the extreme data points to their respective hinges. (Reproduced with permission from Williamson, Parker, and Kendrick, *Annals of Internal Medicine*, 1989;110:917.)

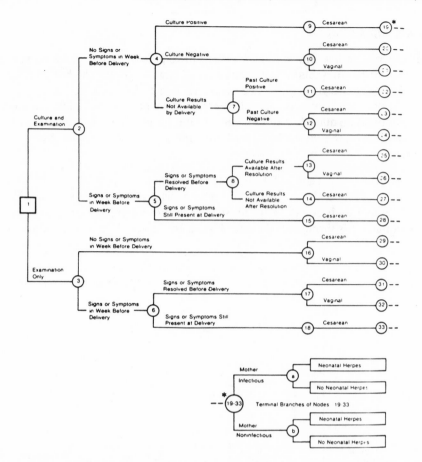

Figure 14-8 Graphic representation of a complex decision tree. The question addressed in the decision analysis is whether to manage recurrent genital herpes during pregnancy by weekly culture and examination or by examination only. The lower right portion of the figure depicts a subtree that represents the outcome at each of the nodes numbered 19–33. (Reproduced with permission from Binkin and Koplan, *Medical Decision Making,* 1989;9:226.)

three categories of quality—poor, good, and best quality. The plots are constructed as follows. Within each quality group, the relative risk estimates are ranked from highest to lowest. The median value (the value with an equal number of estimates of relative risk above and below it) is determined. This value is drawn as the horizontal line inside the box. The middle 50% of relative risk values are determined by counting up and down from the median. These values define the ends of the box. A vertical line is drawn from the middle of each of the crossbars of the box to the most extreme upper and lower values of the relative risk.

14.5 GRAPHICAL PRESENTATION OF THE RESULTS OF DECISION ANALYSIS AND COST-EFFECTIVENESS ANALYSIS

14.5.1 Showing the Decision Tree

The decision tree is virtually always presented graphically. The goal should be presentation of sufficient detail so that the reader can understand the main comparisons. Most journals are not able to publish complex decisions trees in all of their complexity, and some simplification will be necessary. It may be necessary to break the decision down into several parts.

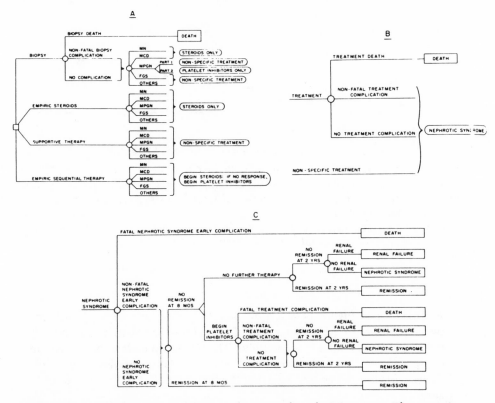

Figure 14-9 Graphic representation of a complex decision tree. The question addressed in the decision analysis is whether to manage adults with idiopathic nephrotic syndrome by biopsy, empiric steroids, supportive therapy, or empiric sequential therapy. The tree is subdivided into three parts that represent events occurring before (panel A), during (panel B), and after (panel C) therapy.

Key: MN = membranous nephropathy; MCD = minimal change disease; MPGN = membranous glomerulonephritis; FGS = focal glomerulosclerosis; OTHERS = other histopathologies. (Reproduced with permission from Levey et al., *Annals of Internal Medicine,* 1987;107:700.)

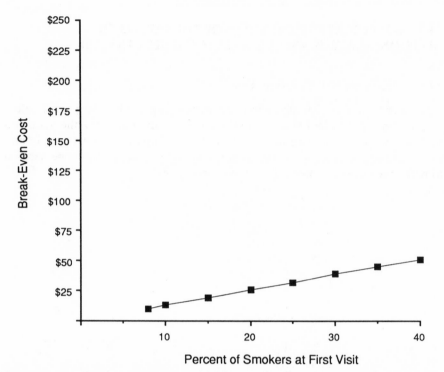

Figure 14-10   Graphic representation of the results of a threshold analysis. The line shows how many dollars could be invested in a smoking cessation program during pregnancy and still "break-even" in terms of averted costs of care for sick newborns and maternal complications according to the prevalence of smoking in pregnant women at the time of the first prenatal visit. (Reproduced with permission from Shipp et al., *American Journal of Public Health,* 1992;82:388.)

EXAMPLE: Figure 14-8 is a decision tree from an analysis comparing two approaches to the management of pregnant women with recurrent genital herpes (Binkin and Koplan 1989). The tree is too large to be printed on one page. The terminal nodes depict identical outcomes. The figure thus shows parts of the tree that repeat at the nodes numbered 19–33 in a subfigure.

An even more complicated decision tree, broken into three parts, is shown in Figure 14-9. Levey et al. (1987) did a decision analysis examining four different strategies for managing adults with idiopathic nephrotic syndrome. The decision problem is complex because it compares four different management strategies and because the outcomes of the strategies are complex. The tree is broken into three separate parts, for events that occur before, during, and after treatment.

14.5.2 Results of Sensitivity Analysis

The results of two-way and three-way sensitivity analyses are best presented graphically. Chapter 13 gave examples of graphic presentation of two-way and three-way sensitivity analysis from the study by Heckerling and Verp 1991) comparing amniocentesis and chorionic villus sampling.

When threshold analysis is done, graphic presentation of the results helps the reader interpret the results.

EXAMPLE: Figure 14-10 shows the results of a threshold analysis in which the "break-even" cost of a smoking cessation program during pregnancy was estimated according to the prevalence of smoking at the time of the first prenatal visit. At each estimate of the prevalence of smoking, the line shows the number of dollars that could be invested in a program of smoking cessation and break-even in terms of the averted costs of care for sick newborns and mothers. The higher the prevalence of smoking, the greater the number of dollars that could be invested and still break-even.

15

Limitations

This book shows how to maximize the usefulness of meta-analysis, decision analysis, and cost-effectiveness analysis by ensuring that studies that use the methods are rigorous. Throughout the book, important controversies about the statistical and mathematical theories that underlie the methods are highlighted, and key assumptions are identified. The main problems with application of the methods and some ways to overcome these problems are presented. The book advocates the three methods. Having taken a position of advocacy in the first 14 chapters, this chapter identifies some serious, sometimes crippling, limitations of the methods.

Section 15.1 discusses the main criticisms of meta-analysis, emphasizing the particular limitations of meta-analysis of nonexperimental data. Section 15.2 discusses framing and the implications of the existence of framing effects for conclusions based on decision analysis. Section 15.3 discusses in more detail the reliance on charge data as a proxy for cost in cost-effectiveness analysis; the overall quality of cost-effectiveness analysis in medical applications is discussed. Section 15.4 shows how the use of life expectancy as a measure of effectiveness in decision analysis and cost-effectiveness analysis has some unappreciated and unpleasant consequences. Section 15.5 discusses the criticisms of quality adjustment and the use of quality-adjusted life years that are based on morals and ethics. Section 15.6 summarizes the situations in which each method is most likely to be useful and the situations in which each method may be misleading or wrong.

15.1 META-ANALYSIS

15.1.2 Overview

Some of the concerns about meta-analysis are broad concerns based on philosophy. Meta-analysis has been subjected to formal evaluations, and the result of these formal evaluations are not all favorable. Moreover, the inability of statistical aggregation methods to overcome problems of bias and uncontrolled confounding is a major problem of meta-analysis of nonexperimental studies.

15.1.2 Broad Concerns About Meta-Analysis

Although the use of meta-analysis is growing rapidly, enthusiasm about it is not universal. Eysenck (1978), in an often-quoted letter titled "an exercise in megasilliness," said about meta-analysis: "'Garbage in–garbage out' is a well-known axiom of computer specialists; it applies here with equal force."

Commenting on the use of meta-analysis in the social sciences, Cook and Leviton (1980) point out that the "apparent objectivity, precision and scientism of meta-analysis lend social credibility that may be built on procedural invalidity."

Wachter (1988) identifies four reasons for skepticism about meta-analysis. The first reason is the "garbage-in–garbage-out" criticism articulated by Eysenck. The second reason is the concern, more prominent in the social sciences than in medicine, that the measurements of outcome in studies that are combined in a meta-analysis are often different measures and that meta-analysis "compares apples and oranges." The third reason for concern about meta-analysis is that the study of previous studies is being reduced to the routinized task of coding relegated to the research assistant. Last, skeptics are worried that meta-analysis is part of a world in which "bad science drives out good by weight of numbers."

Of most importance is the last concern, the concern that the weight of numbers and the precision of meta-analysis drive out good science. Meta-analysis appears to be very precise and can carry considerable weight in policy discussions. The certainty with which the result of a meta-analysis is presented may dissuade persons from doing original studies on the same topic. The possibility that the results of a meta-analysis may limit the availability of funds for additional studies must be taken seriously.

EXAMPLE: Following publication of the results of the meta-analysis of clinical trials of adjuvant chemotherapy for node-negative breast cancer that was described in Chapter 1 (Early Breast Cancer Trialists' Collaborative Group 1988), the National Cancer Institute issued a "Clinical Alert" to all physicians informing them of the findings of the analysis and urging immediate changes in clinical practice based on the results. In this situation, a new clinical trial of the use of adjuvant chemotherapy would almost certainly be turned down by a local ethics committee, and funding is highly unlikely to be available readily.

15.1.3 Formal Evaluations of Meta-Analysis

The quality of meta-analysis, its reproducibility, and its correlation with the results of "definitive" studies have all been formally evaluated. The results of these evaluations give a mixed picture of meta-analysis.

Sacks et al. (1987) evaluated the quality of 86 published meta-analyses. They found that only 24 of 86 meta-analyses addressed all six major areas that were considered to measure the quality of meta-analysis. Table 15-1 shows the number and the percentage of the 86 studies that were considered adequate for each of the quality features. For 18 of the 23 features, more than 50% of the meta-analyses were rated as inadequate.

Chalmers et al. (1987) compared the results of a meta-analysis of multiple undersized randomized trials with the results of a large "gold standard" trial for three separate topics. Table 15-2 describes the number of subjects in the meta-

Table 15-1 Adequacy of 86 meta-analyses published in medical journals in 1966–1986.

| | Studies Considered Adequate | |
Criterion	N	%
Design		
Protocol	6	7.0
Literature search	30	34.9
List of trials analyzed	81	94.2
Log of rejected trials	11	12.8
Treatment assignment	59	68.6
Ranges of patients	19	22.1
Ranges of treatment	39	45.3
Ranges of diagnosis	34	39.5
Combinability		
Criteria	39	45.3
Measurement	20	23.3
Control of bias		
Selection bias	0	0.0
Data-extraction bias	0	0.0
Interobserver agreement	4	4.7
Source of support	28	32.6
Sensitivity analysis		
Quality assessment	16	18.6
Varying methods	14	16.3
Publication bias	2	2.3
Application of results		
Caveats stated	66	76.7
Economic impact	1	1.2

Source: Sacks et al. (1987).

Table 15-2 For three topics, summary of odds ratio from a meta-analysis of multiple small randomized clinical trials with the odds ratio from a large "gold standard" trial

Topic	Endpoint	Number of Studies in Meta-Analysis	Patients		Odds Ratio	
			Meta-Analysis	"Gold Standard" Trial	Meta-Analysis	"Gold Standard" Trial
Effect of intravenous beta blockers after acute myocardial infarction[a]	Short-term mortality	24	4,408	16,027	0.94	0.85
Intravenous streptokinase after acute myocardial infarction	Reinfarction	10	4,947	11,712	0.7	1.9
	Mortality	11	5,268	11,712	0.80	0.81
Phenobarbital for prevention of intracranial hemorrhage in the newborn	Severe ICH[b]	7	413	280	0.6	2.1
	All ICH[b]	7	413	280	0.6	2.1

[a]Data on results of the meta-analysis of intravenous beta blockers and short-term mortality are from the publication of Yusuf et al. (1985B).
[b]Intracranial hemorrhage.
Source: Chalmers et al. (1987).

analysis and in the gold standard study and shows the odds ratios for treatment as derived from the meta-analyses and the gold standard studies. Table 15-3 shows whether the 95% confidence intervals from the meta-analysis and from the gold standard study did or did not exclude 1.0.

For all three topics, the results of the meta-analysis and the "gold standard" study were somewhat discrepant. Thus, in a meta-analysis of 24 studies of the effect of intravenous beta blockers in myocardial infarction that was published before the gold standard study, Yusuf et al. (1985B) reported an odds ratio for short-term mortality in the treatment group of 0.94 with a 95% confidence limit that *included* 1.0, whereas the "gold standard" study (ISSI-1 Collaborative Group 1986) reported an odds ratio of 0.85 with a 95% confidence limit that *excluded* 1.0. The meta-analysis of phenobarbital and intracranial hemorrhage in neonates showed a *lower* risk of both severe hemorrhage and all hemorrhage in the treatment group, whereas the gold standard study (Kuban et al. 1986) found a *higher* risk of both severe hemorrhage and all hemorrhage in treated neonates. If a 95% confidence interval excluding 1.0 is used as a test of statistical significance, the *lower* risk of all hemorrhage based on the meta-analysis was statistically significant; the *higher* risk of all hemorrhage in the gold standard trial was also statistically significant. The results of the meta-analysis and the gold standard trial

Table 15-3　For three topics, summary of comparison of 95% confidence interval from meta-analysis of multiple small randomized clinical trials with the confidence interval from a large "gold standard" trial

| | | 95% C.I. − 1.0 | |
| | | Meta- | "Gold Standard" |
Topic	Endpoint	Analysis	Trial
Effect of intravenous beta blockers after acute myocardial infarction[a]	Short-term mortality	No	Yes
Intravenous streptokinase after acute myocardial infarction	Reinfarction	No	Yes[b]
	Mortality	Yes	Yes
Phenobarbital for prevention of intracranial hemorrhage in the newborn	Severe ICH[c]	No	No[b]
	All ICH[c]	Yes	Yes[b]

[a]Data on the results of the meta-analysis of intravenous beta blockers and short-term mortality are from the publication of Yusuf et al. (1985B).
[b]Odds ratios are in the opposite direction.
[c]Intracranial hemorrhage.
Source: Chalmers et al. (1987).

(GISSI 1986) for intravenous streptokinase were essentially identical when mortality was used as the endpoint. With myocardial reinfarction as an endpoint, the odds ratio for treatment was 0.7 in the meta-analysis and 1.9 in the gold standard trial, although the 95% confidence intervals did not exclude 1.0 for the meta-analysis.

Chalmers et al. (1987) also compared the results of independent replications of meta-analyses of the same topic for 18 different topics. Table 15-4 shows the details of their comparisons for the topics where the same endpoint was used in all of the meta-analyses, and Table 15-5 summarizes the comparison. For 12 of the 18 topics, all of the replicate meta-analyses agreed both on the direction of the effect of the intervention and on whether or not the effect was statistically significant at probability level of 0.05. For 3 of the 18 topics, at least one of the replicates found an effect opposite in direction to another replicate. The findings with regard to the replicability of meta-analysis are more encouraging than the findings comparing meta-analysis with definitive studies, but the replicability of meta-analysis is by no means proven. The replicability of meta-analysis of nonexperimental studies is unstudied.

15.1.4　Particular Problems with the Meta-Analysis of Nonexperimental Studies

Meta-analysis started as a method to summarize the results of experimental studies. It was a particular response to problems arising from the conduct of small randomized trials that individually showed no effect of the intervention but, when combined, were statistically significant. In experimental studies, randomization in theory eliminates bias and confounding, and the measure of the effect of the intervention derived from a randomized trial is considered an unbiased and uncon-

Table 15-4 For 18 topics where two or more meta-analyses have been done, number that agree on direction of effect and statistical significance, on direction of effect but not significance, and number that disagree on direction of effect

Topic	Number of Meta-Analyses of this Topic	Number That Agree on Both Direction of Effect and Significance	Number That Agree on Direction of Effect but Not Significance	Disagree on Direction[a]
Long-term beta blockers post-MI	6	6	0	0
Short-term beta blockers post-MI	4	4	0	0
Intravenous streptokinase post-MI	3	3	0	0
Aspirin post-MI	3	3	0	0
Psychoeducation intervention	2	2	0	0
Patient education	2	2	0	0
Nicotine chewing gum in clinics	2	2	0	0
Nicotine chewing gum in practice	2	2	0	0
Prevention of venous thromboembolism	2	2	0	0
Stimulant therapy for hyperactivity	2	2	0	0
Psychotherapy	5	4	0	1
Steroids in alcohol hepatitis	3	1	2	0
Antidepressant drugs	3	2	1	0
Single-dose TMP-SMZ for UTI	3	1	1	1
Single-dose amoxicillin for UTI	3	0	3	0
Anticoagulants post-MI	3	2	0	1
Radiotherapy after mastectomy	2	2	0	0
Steroids and peptic ulcer	2	2	0	0

[a]Includes studies where one meta-analysis favored treatment or control and the other found no effect.
Source: Chalmers et al. (1987).

Table 15-5 Summary of data on replicates of meta-analysis shown in Table 15-4

Result	Number of Topics
All replicates agree on both direction of effect and statistical significance	12
At least some replicates disagree on direction of effect	3
Other	3

founded measure of the effect of the intervention. Having (theoretically) eliminated bias and confounding as an explanation for the observed differences between the treatment and control groups, the critical remaining issue in a meta-analysis of experimental studies is comprehensive identification of information pertaining to the question addressed in the meta-analysis and unbiased extraction of pertinent data from all of the relevant studies.

Meta-analysis of nonexperimental studies has followed on the heels of meta-analysis of experimental studies without much critical writing and absent formal consideration of the differences in the most important problems confronting the meta-analysis of nonexperimental and experimental studies. Meta-analysis of nonexperimental studies seems to be done most often when the results of the individual studies are contradictory, not when they are individually too small to provide an answer to the question addressed.

EXAMPLE: Four different meta-analyses of estrogen replacement therapy and breast cancer have been published (Armstrong 1988; Dupont and Page 1991; Steinberg et al. 1991; Sillero-Arenas et al. 1992). These meta-analyses were not undertaken to remedy problems due to the small size of studies of breast cancer and estrogen replacement therapy. In fact, each of three different studies of this topic had more than 1000 cases of breast cancer; one study had over 5000 cases; and the total number of cases of breast cancer in all of the studies of the topic combined is more than 10,000. Meta-analysis was used for this topic in an effort to resolve the contradictions between study results.

Unfortunately, meta-analysis is weakest and most controversial when studies disagree and there is heterogeneity. Use of a random-effects model will better reflect the uncertainty in the body of data on the topic, but statistical analysis alone will not make the results of contradictory studies agree.

The emphasis on the mechanics of combining estimates of effect from nonexperimental studies of different designs has diverted attention from the more critical question of how to handle studies where there is concern about bias and uncontrolled confounding. When the results of an individual study *may* be due to bias or there *may* be uncontrolled confounding, it is usually impossible entirely to rule out bias and confounding as an explanation for the results of that study. In a group of studies where some studies are "positive" and some are "negative," the possibility that the positive ones are due to bias or uncontrolled confounding and the negative ones are free of bias and confounding (or vice versa) is difficult, if not impossible, to rule out.

Even when the results of virtually all of the studies are consistent, it is still possible that all of them suffer from the same bias.

EXAMPLE: It is argued that studies of lung cancer and passive smoking are flawed by misclassification of active smokers as non-smokers due to underreporting of active smoking. If persons who are truly active smokers are also more likely to be exposed to passive smoke and assuming that active smoking is a very strong risk factor for lung cancer, underreporting of active smoking will bias estimates of the relative risk of lung cancer in relation to passive smoking even if the underreporting is nondifferential. Because it is impossible to validate statements about past active smoking, even with physiologic measures of smoke exposure, this source of bias cannot be eliminated. Underreporting of active smoking may be a relatively universal problem, and the fact that almost all studies of lung cancer and passive smoking are "positive" does not eliminate the possibility that this bias may explain the results of a meta-analysis, since the statistical aggregation of data does not address the possibility of this bias.

15.2 DECISION ANALYSIS

15.2.1 Overview

Decision analysis rests on the mathematical theory of choice. The existence of framing effects raises serious questions about certain aspects of this fundamental theory. There are no formal evaluations of the reproducibility or quality of decision analysis, and the very absence of critical evaluation of the method is a limitation.

15.2.2 Framing Effects and the Philosophy of Choice

Tversky and Kahnemann (1981) have shown that seemingly inconsequential changes in the formulation of choice problems can cause radical shifts in measured preferences.

> *EXAMPLE:* Two groups of students in a classroom setting were given the following problem:

> Imagine that the United States is preparing for the outbreak of an unusual Asian disease, which is expected to kill 600 people. Two alternative programs to combat the disease have been proposed.

The two sets of students were asked to choose between two programs. The first set of students ($n = 152$) was given the choice between the following two programs:

If Program A is adopted, 200 people will be saved.

If Program B is adopted, there is a 1/3 probability that 600 people will be saved and 2/3 probability that no people will be saved.

The second set of students ($n = 155$) was given the choice between the following two different programs:

If Program C is adopted 400 people will die.

If Program D is adopted there is 1/3 probability that nobody will die and 2/3 probability than 600 people will die.

In the choice between Program A and Program B, 72% of students chose Program A. In the choice between Program C and Program D, 78% of students chose program D. The only real difference between the two sets of programs is that the first involves the number of lives saved and the second the number of lives lost. Through this and other work, Tversky and Kahnemann (1981) show a consistent pattern in which choices involving certain gains are risk averse and choice involving certain losses are often risk taking.

The influence of the manner in which a problem is presented on choice is an example of a "framing" effect. If changes in "frame" can cause large changes in individual preferences, then choices cannot be assumed to be fixed and stable. If choices are not fixed and stable, then the general theory that is the basis for decision analysis must be questioned (Tversky and Kahnemann 1981).

15.2.3   Formal Evaluations of Decision Analysis

Formal evaluations of the quality of decision analysis studies have not been done. Replications of decision analyses could not be identified. Financial support to carry out studies of quality and reliability is limited, and this probably accounts for the paucity of work on evaluation of the methods. The absence of information proving the quality and the reliability of decision analysis seriously limits the method.

15.3   COST-EFFECTIVENESS ANALYSIS

15.3.1   Overview

The frequent substitution of charge data for data on cost in spite of the concerns about this practice casts doubt on many cost-effectiveness analyses. Formal evaluation of the quality of cost-effectiveness analysis in medicine has been very limited. The evaluation that has been done is disheartening.

15.3.2   Using Charge as a Proxy for Cost

It is axiomatic that a valid cost-effectiveness analysis must correctly assess cost, yet true measures of cost are almost never used in cost-effectiveness analyses in medicine. Table 15-6 shows the source of cost information for six cost-effectiveness analyses that were published in 1987 or later in either of two prestigious medical journals, the *New England Journal of Medicine* or the *Journal of the American Medical Association*. Each analysis required data on two to five costs. Of the six cost-effectiveness analyses, only one (Castellano and Nettleman 1991) used true cost data for every cost item in the analysis. Three analyses (Arevalo and Washington, 1988; Goldman et al. 1988; Kinosian and Eisenberg, 1988) used charge, payment, or reimbursement data for all items. Two (Goldman et al. 1991; Oster and Epstein 1987) used a mix of cost data and charge, payment, or reimbursement data.

When the perspective of an analysis is a program perspective, the use of charge or payment data may be appropriate. None of the six analyses listed in Table 15-6 took a program perspective, and the use of charge, payment, and reimbursement

Table 15-6 Source of data on "cost" for six published cost-effectiveness analyses described in this book and published in prestigious journals[a]

Reference	Topic	Perspective of Analysis	Description of Costs	Source of Data on "Cost"
Arevalo and Washington (1988)	Hepatitis B immunization	Societal	Screening test for HBsAg	Average charge in several local laboratories
			Hepatitis B immunization	Uncertain
			Hospital cost for liver cancer	Hospital charge
			Care for chronic liver disease	Hospital charge
Castellano and Nettleman (1991)	PCP prophylaxis	Societal	Hospitalization for PCP	Cost
			Aerosolized pentamidine	Cost
Goldman et al. (1991)	Cholesterol treatment	Societal	HMG-CoA reductase inhibitor	Average charge in 10 regional pharmacies
			Hospitalization for CHD	Cost
			Physician visits for monitoring	Uncertain
			Laboratory tests	Uncertain
Goldman et al. (1988)	Beta blocker treatment	Societal	Cost of beta blockers	Average charge in 10 regional pharmacies
			Hospitalization	Uncertain
Oster and Epstein (1987)	Cholesterol treatment	Societal	Cholestyramine	Average retail price in pharmacies based on national survey
			Office visits for therapy and side effects	Blue Shield reimbursement rate in Massachusetts
			CHD	Cost
			Induced costs	Expenditures for health care in a national survey
Kinosian and Eisenberg (1987)	Cholesterol treatment	Societal	Cholestyramine, colestipol	Price in pharmacies
			Oat bran	Uncertain
			Office visits, laboratory tests	Payments for services
			Hospitalization	DRG payment

[a]*New England Journal of Medicine* or *Journal of the American Medical Association.*

Key: HBsAG = hepatitis B surface antigen; PCP = pneumocystis carinii pneumonia; HMG-CoA reductase inhibitor = 3-hydroxy-3-methylglutaryl coenzyme A reductase inhibitor; CHD = coronary heart disease

Table 15-7   Various estimates of the "cost" of pneumococcal vaccination used in cost-effectiveness analysis of pneumococcal vaccination in persons 65 years and older

Description	Amount	Type of "Cost"
Private M.D.		
Charge by manufacturer per dose	$ 4.43	
Physician fee for administering vaccine	$10.22	Charge
Total	$14.65	
Public program		
Charge by manufacturer per dose[a]	$2.22	Resource consumption
Administering vaccine[b]	$1.58	Under specified conditions
Total	$3.80	
Medicare		
Payment by carrier	$9.60	Payment
Total	$9.60	

[a]Reflects volume discount.
[b]In a program setting.
Source: Sisk and Riegelman (1986).

data cannot be justified based on the perspective. In general, substitution of charge for cost data is based on convenience and the difficulty of measuring cost.

The use of charge data as a substitute for cost can lead to unwarranted conclusions about efficiency (Finkler 1982). Thus, estimated "savings" may not materialize if charges are used in place of costs, since in fixed economic systems, charges can be changed at will.

A cost-effectiveness analysis may yield different conclusions depending on whether data on charge or data on payment are used.

EXAMPLE: In the cost-effectiveness analysis of pneumococcal vaccine for persons over 65 done by Sisk and Riegelman (1986) that was discussed briefly in Chapter 1, three estimates of the "cost" of the vaccine were used. These estimates are described in Table 15-7. The first estimate is an estimate of the amount charged by the manufacturer to the physician or a pharmacy purchasing a small quantity of vaccine and the amount charged to patients to administer the vaccine. This amount, $14.65, is cost measured as charge. The second estimate of cost is for a hypothetical public program. It includes the amount charged by the manufacturer for vaccine purchased in bulk quantities and the cost of administering the vaccine in a program setting. This amount, $3.80, is very close to being a measure of true cost, or resource consumption. The third amount is the amount that Medicare reimburses physicians for administering the vaccine to a Medicare beneficiary. This amount, $9.60, is cost measured as payment.

Table 15-8 presents the results of the Sisk and Riegelman (1986) cost-effectiveness analysis when based on a cost estimate of $3.80, which is the estimate that approximates true cost, and on the estimate of $14.65, which an estimate based on charge. Under all of the reasonable assumptions about the efficacy of the vaccine, vaccination is cost saving if the cost of vaccine is $3.80. Since an intervention that is cost saving is cost effective, vaccination is cost effective when based on the true cost of vaccine. Depending on assumptions, the cost per year of life saved ranges from $0.84 to $7.81 when the cost estimate of $14.65 is used in the analysis. In this case, the question of whether or not the vaccination is "cost effective" is dependent on an assessment of the value of a year of life. In this country, a year of life would probably be judged to be worth an investment of $7.81, but this might not be the case in all situations.

The persistence of the practice of using charge data in place of cost data in cost-effectiveness analysis subjects these analyses to well-founded global criticisms about adherence to the principles of economics. The use of charge data in place of cost is a serious threat to the acceptance of the results of cost-effectiveness analysis by both the uniformed and, more importantly, by experts in economics.

15.3.3 Formal Evaluations of Cost-Effectiveness Analysis

Udvarhelyi et al. (1992) did a formal evaluation of the quality of cost-effectiveness analyses published in the medical literature. They identified 77 articles using cost-effectiveness or cost-benefit analysis published in general medical, general surgical, and medical subspecialty journals from 1978 to 1980 and from 1981 to 1987. They assessed the use and reporting of six fundamental principles of cost-effectiveness analysis. The results of their evaluation are summarized in Table 15-9.

Only 3 of the 77 articles on cost-effectiveness analysis adhered to all six fundamental principles of cost-effectiveness analysis. In both periods, more than 50% of articles failed to make an explicit statement of the perspective of the analysis, failed to include the costs of side effects, averted costs, and induced costs, failed to use sensitivity analysis, or failed to use a preferred measure of cost effectiveness.

Table 15-8 Net medical cost per year of life saved for pneumococcal vaccination of persons 65 years or older, for several assumptions about percentage of pneumococcal pneumonia, duration of immunity, and cost per vaccination

	Cost per Vaccination[a]	
Assumption	$3.80 (true cost)	$14.65 (charge)
15% of pneumonia is pneumococcal; immunity lasts 8 years	(—)[b]	(—)[b]
10% of pneumonia is pneumococcal; immunity lasts 8 years	(—)[b]	$0.84
15% of pneumonia is pneumococcal; immunity lasts 3 years	(—)[b]	$4.33
10% of pneumonia is pneumococcal; immunity lasts 3 years	(—)[b]	$7.81

[a]In 1983 dollars.
[b]Cost saving.
Source: Sisk and Riegelman (1986).

Table 15-9 Nonadherence to six principles of cost-effectiveness analysis for 77 articles published in general medical, general surgical, and medical subspecialty journals

	Articles Not Adhering to Principle			
	1978–1980 (N = 31)		1985–1987 (N = 46)	
Principle	N	%	N	%
Perspective explicitly stated	23	74.2	40	87.0
Benefits explicitly stated	5	16.1	8	17.4
Costs				
Cost data provided	11	35.5	2	4.3
Program or treatment costs included	6	19.4	4	8.7
Side effect or morbidity costs included	24	77.4	30	65.2
Averted costs included	23	74.2	35	76.1
Induced costs included	30	96.8	44	95.7
Timing				
Discounting used if timing of costs and benefits are different	9/15	60.0	9/14	42.9
Sensitivity analysis done	24	77.4	30	65.2
Preferred summary measurement used	24	77.4	34	73.9

Source: Udvarhelyi et al. (1992).

It is unusual for more than one cost-effectiveness analysis of the same topic to be done. Whether the rarity of replicate studies of cost-effectiveness analysis reflects an unwillingness of journals to publish replications or of investigators to undertake them is uncertain. The difficulty of carrying out cost-effectiveness analysis, the small number of economists trained in health issues, and the lack of funding for cost-effectiveness analysis probably all contribute to the rarity of replicate studies. Reproducibility is a cornerstone of the scientific method, and the absence of data to establish the reproducibility of cost-effectiveness analysis is a serious limitation of the method.

15.4 PROBLEMS WITH LIFE EXPECTANCY AS AN OUTCOME MEASURE

Life expectancy is often the outcome measure of decision analysis and cost-effectiveness analysis. Life expectancy in the absence of an intervention is a function of current age and sex and, in most populations, of race. When life expectancy is short, the estimated effect of an intervention will be small even if the intervention has the same absolute effect on mortality per year in every group.

EXAMPLE: Table 15-10 shows estimated life expectancy for 45-year-old white women, 45-year-old black men, and 90-year-old white women. The table also shows the average mortality rate based on these estimates of life expectancy along with the estimated gain in life expectancy from an

Table 15-10 Estimated gain in life expectancy for three groups whose baseline life expectancies differ

Group	Life Expectancy (years)	Average Mortality Rate per Year	Gain in Life Expectancy from Hypothetical Intervention (Years)[a]
45-year-old white women	37.8	0.026	+1.40
45-year-old black man	27.8	0.036	+0.80
90-year-old white women	5.2	0.192	+0.02

[a]For an intervention that decreases mortality by 0.001 per year.

intervention that decreases overall mortality by 0.001 per year in each group. These calculations were carried out as described in Section 9.4.2 using the DEALE.

The gain in life expectancy from the hypothetical intervention is 0.02 year in 90-year-old white women, 0.80 year in 45-year-old black men, and 1.40 years in 45-year-old white women. In each group, 1 of 1000 persons alive at the beginning of the year are "saved" by the intervention. It is only *cumulatively,* and because of the lower likelihood of death from other causes, that more persons are saved and more life years are gained in the younger white women.

In general, when life expectancy is used as the measure of effectiveness, an intervention that prolongs life will have the smallest effect on gain in life expectancy in the group with the shortest life expectancy. When a cost-effectiveness analysis is done, the cost per year of life gained will be greatest in the group with the shortest life expectancy in the absence of the intervention.

EXAMPLE: Assume that the cost of the intervention described in the preceding example is $100,000. The alternative intervention is to do nothing, which costs nothing. Assume that the cost of the intervention is a one-time cost and thus discounting of costs does not affect the estimate of cost effectiveness. Table 15-11 shows the estimated cost per year of life gained for the hypothetical intervention for middle-aged white women, middle-aged black men, and elderly white women. For 45-year-old white women, the cost of the intervention is only $71,429 per year of life gained; for 45-year-old black men, it is $125,000 per year of life gained; for 90-year-old white women, it is $5,000,000 per year of life.

When an analysis poses a choice between alternative therapies for a defined group or for an individual, this problem with life expectancy does not affect decision analysis and cost-effectiveness analysis. When the analysis is used to guide choices between an intervention for a person with a short life expectancy and the same intervention for a person with a longer life expectancy, decision analysis and cost-effectiveness analysis will "discriminate" against the group whose life expectancy is shortest (Harris 1987).

Table 15-11 Estimated gain in life expectancy and cost per year of life gained for three groups whose baseline life expectancies differ

Group	Gain in Life Expectancy (years)[a]	Cost per Year in Life Gained
45-year-old white women	+1.40	$ 71,429
45-year-old black men	+0.80	125,000
90-year-old white women	+0.02	5,000,000

[a]For an intervention that decreases mortality by 0.001 per year in each group.

15.5 VALUES AND ETHICS AND THE QUALITY-ADJUSTED LIFE YEAR

The use of quality-adjusted life years as the measure of effectiveness in decision analysis and cost-effectiveness analysis has stirred great controversy. The controversy generally does not focus on the formidable technical and practical problems with the measurement of preferences that were described in Chapter 11. It centers on issues of values, ethics, and morals.

Quality-adjusted life years can be used in two ways (Harris 1987). First, they can be used to determine which of alternative treatments should be given to a particular patient or which of alternative treatments to manage the same condition should be favored. Second, quality-adjusted life years can be used to determine which patients to treat or which conditions to give priority in the allocation of health resources.

EXAMPLES: Weinstein and Stason's (1985) analysis that compared the quality-adjusted life expectancy of CABG and medical management of patients with coronary artery disease is an example of the use of quality-adjusted life years in the first way, to inform decisions about how to manage coronary heart disease in individual patients. It is logical to extend this use of quality-adjusted life years to decisions about which treatment should be favored programmatically.

Williams (1985) did an analysis comparing the cost per quality-adjusted life year of coronary artery bypass graft with the cost per quality-adjusted life year of renal transplantation and concluded that bypass grafting for left main coronary artery disease and triple vessel disease should be funded before renal transplantation. The state of Oregon developed a method for estimating the cost per unit of quality-adjusted life for a large group of medical services and interventions and attempted to use these ratings to decide which services would be covered by a state-funded insurance program for the medically underserved. These examples are examples of the use of quality-adjusted life years to make choices among treatments and to set priorities for expenditures in a population.

The use of quality-adjusted life years to help guide choices between alternative treatments for a single patient and investments of society in one intervention in preference to another intervention for the same condition generally is held to be useful (Harris 1987; Smith 1987). However, even considering this generally accepted use of quality-adjusted life years, there are arguments (Harris 1987; Smith 1987) that assigning different values to similar durations of life imposes a judgment about the value of life with which affected individuals might not concur. In addition, some reject entirely the notion of quality of life as a consideration in decision making. Rawles (1989), for example, states that giving a quality of life score of 1 to the absence of suffering equates the value of life with the absence of disability and distress and undervalues existence itself.

The use of estimates of quality-adjusted life years to make decisions about how to determine which patients to treat and how to establish priorities for treatments for different conditions is the most controversial, as well as the most common use of quality-adjusted life years. Using quality-adjusted life years in this way has been called "positively dangerous and morally indefensible" by one author (Harris 1987) and based on "false premises, faulty reasoning, and unjust principles" by another (Rawles 1989).

Concerns about the justness and the morality of the use of quality-adjusted life years to determine who to treat and treatment priorities are mostly concerns made in the context of their use in cost-effectiveness analysis. Drummond (1987) points out that investing in the interventions that have the lowest cost per quality-adjusted life year ignores the principle of equity. That is, cost-effectiveness analysis produces the largest improvement in health status given a fixed set of resources but fails to take into account the needs of disadvantaged groups. Second, use of quality-adjusted life years to decide who to treat or what to pay for ignores what might be the choices of individuals. Thus, it abrogates the ethical principle of autonomy, which is generally most important for individual patients, to the principle of justice or fairness, which is generally most important for a community (LaPuma and Lawlor 1990). Third, the use of quality-adjusted life years discriminates against the aged, for reasons detailed in the preceding section, and against the disabled, because their quality of life is downweighted in an analysis (Harris 1987).

Concrete examples of irrationality, logical inconsistencies, and complex moral dilemmas created by the use of quality-adjusted life years in deciding who to treat or which treatment to give priority are easy to find in the critical literature on the ethics and morality of use of quality-adjusted life years (Harris 1987; Rawles 1989). Rawles (1989), for example, points out that the best value for money considering quality-adjusted life years is to turn off the life support machinery for an unconscious patient, whose quality of life is rated negatively and whose treatment is very costly. He shows that the number of quality-adjusted life years gained for two treatments, one of which prolongs the lives of 10 people by 1 year and the other of which saves the life of the 1 person in 10 who die 10 years prematurely, is identical. If the two treatments had the same cost, there would be no reason to favor one over the other considering their cost effectiveness. He argues that most people would choose to save the life of the one person who otherwise would have died early, on grounds of equity.

The counterarguments to the use of quality-adjusted life years to decide who to treat and how to allocate resources focus on the seriousness of the dilemma of allocating resources and the lack of rational alternatives to cost-effectiveness analysis (Danford 1990; Kaplan and Ganiats 1990). "Rationing" is occurring already, it is argued, and cost-effectiveness analysis simply makes explicit the basis for decisions about how to allocate resources.

The resolution of these arguments is not within the scope of this book. Users of the methods need to appreciate the arguments and to understand their basis.

15.6 SITUATIONS WHERE THE METHODS ARE MOST AND LEAST USEFUL

Meta-analysis is most useful for aggregating the results of randomized trials when the results of the trials are generally consistent. When registries of all trials that have ever been done exist, the usefulness of meta-analysis of experimental studies is especially high.

Applied to nonexperimental studies, meta-analysis is most useful when there are many small studies that are consistent in their findings and as a way of identifying and exploring the reasons for heterogeneity in study results. Meta-analysis is least useful when it used solely to derive a single estimate of effect size. Meta-analysis will not resolve the inconsistencies among the results of nonexperimental studies when the results of the individual studies are in conflict. Meta-analysis does not eliminate bias in individual studies.

Both decision analysis and cost-effectiveness analysis are most useful when they are used to try to decide which of two or more alternative treatments to use for a patient or which alternative to prefer for the same condition. Both methods are most useful when information on the probabilities to be estimated is obtained from systematic review of the medical literature.

Cost-effectiveness analysis seems unlikely to solve the health funding crisis that grips our country, since it fails to resolve the profound moral and ethical dilemmas that are the heart of the crisis.

References

Antiplatelet Trialists' Collaboration: Secondary prevention of vascular disease by prolonged antiplatelet treatment. *Br Med J* 1988;296:320–332.

Arevalo JA, Washington AE: Cost-effectiveness of prenatal screening and immunization for hepatitis B virus. *JAMA* 1988;259:365–369.

Armitage P, Berry G: *Statistical Methods in Medical Research,* 2nd ed. Oxford, Blackwell Scientific Publications, 1987, pp 409–410.

Armstrong BK: Oestrogen therapy after the menopause—boon or bane? *Med J Aust* 1988;148:213–214.

Bailey KR: Inter-study differences: how should they influence the interpretation and analysis of results? *Stat Med* 1987;6:351–358.

Bayarri MJ: Comment on "Selection models and the file drawer problem." *Stat Science* 1988;3:128–131.

Beck JR, Kassirer JP, Pauker SG: A convenient approximation of life expectancy (the "DEALE"): I. Validation of the method. *Am J Med* 1982;73:883–888.

Beck JR, Pauker SG: The Markov process in medical prognosis. *Med Decis Making* 1983;3:419–458.

Beck JR, Pauker SG, Gottlieb JE, Klein K, Kassirer JP: A convenient approximation of life expectancy (the "DEALE"): II. Use in medical decision-making. *Am J Med* 1982;73:889–897.

Begg CB, Berlin JA: Publication bias: a problem in interpreting medical data. *J Royal Stat Soc A* 1988;151:419–463.

Berlin JA, Laird NM, Sacks HS, Chalmers TC: A comparison of statistical methods for combining event rates from clinical trials. *Stat Med* 1989;8:141–151.

Binkin NJ, Koplan JD: The high cost and low efficacy of weekly viral cultures for pregnant women with recurrent genital herpes: a reappraisal. *Med Decis Making* 1989;9:225–230.

Boyle MH, Torrance GW: Developing multiattribute health indexes. *Med Care* 1984;22:1045–1057.

Boysen G, Nyboe J, Appleyard M, Sorensen PS, Boas J, Somnier F, Jensen G, Schnor P: Stroke incidence and risk factors for stroke in Copenhagen, Denmark. *Stroke* 1988;19:1345–1353.

Bulpitt CJ: Meta-analysis. *Lancet* 1988;ii:93–94.

Bush TL, Barrett-Connor E, Cowan LD, Criqui MH, Wallace RB, Sutchindran CM, Tyroler HA, Rifkind BM: Cardiovascular mortality and noncontraceptive use of estrogen in women: results from the Lipid Research Clinics Program Follow-up Study. *Circulation* 1987;75:1102–1109.

Castellano AR, Nettleman MD: Cost and benefit of secondary prophylaxis for *pneumocystis carinii* pneumonia. *JAMA* 1991;266:820–824.

Centers for Disease Control: Measles—United States, 1989 and first 20 weeks 1990. *MMWR* 1990;39:353–355, 361–363.

Chalmers I, Adams M, Dickersin K, Hetherington J, Tarnow-Mordi W, Meinert C, Tonascia S, Chalmers TC: A cohort study of summary reports of controlled trials. *JAMA* 1990;263:1401–1405.

Chalmers TC, Block JB, Lee S: Controlled studies in clinical cancer research. *N Engl J Med* 1972;287:75–78.

Chalmers TC, Smith H Jr, Blackburn B, Silverman B, Schroeder B, Reitman D, Ambroz A: A method for assessing the quality of a randomized control trial. *Controlled Clin Trials* 1981;2:31–49.

Chalmers TC, Levin H, Sacks HS, Reitmen D, Berrier J, Nagalingam R: Meta-analysis of clinical trials as a scientific discipline: I. Control of bias and comparison with large cooperative trials. *Stat Med* 1987A;6:315–325.

Chalmers TC, Berrier J, Sacks HS, Levin H, Reitmen D, Nagalingam R: Meta-analysis of clinical trials as a scientific discipline: II. Replicate variability and comparison of studies that agree and disagree. *Stat Med* 1987B;6:733–744.

Cochran WG: Problems arising in the analysis of a series of similar experiments. *J Royal Stat Soc B* 1937;4:102–118.

Cochran WG: The combination of estimates from different experiments. *Biometrics* 1954;10:101–129.

Colditz GA, Miller JN, Mosteller F: How study design affects outcomes in comparisons of therapy: I. Medical. *Stat Med* 1989;8:441–454.

Collins R, Yusuf S, Peto R: Overview of randomized trials of diuretics in pregnancy. *Br Med J* 1985;290:17–23.

Cook TD, Leviton LC: Reviewing the literature: a comparison of traditional methods with meta-analysis. *J Personality* 1980;48:449–472.

Critchfield GC, Willard KE: Probabilistic analysis of decision trees using Monte Carlo simulation. *Med Decis Making* 1986;6:85–92.

Danford DA: QALYs: their ethical implications. *JAMA* 1990;264;2503.

Demets DL: Methods for combining randomized clinical trials: strengths and limitations. *Stat Med* 1987;6:341–348.

DerSimonian R, Laird N: Meta-analysis in clinical trials. *Controlled Clin Trials* 1986;7:177–188.

Detsky AS, Naglie IG: A clinician's guide to cost-effectiveness analysis. *Ann Intern Med* 1990;113:147–154.

Devine EC, Cook TD: A meta-analytic analysis of effects of psychoeducational interventions on length of postsurgical hospital stay. *Nursing Res* 1983;32:267–274.

Dickersin K, Hewitt P, Mutch L, Chalmers I, Chalmers TC: Comparison of MEDLINE searching with a perinatal trials database. *Controlled Clin Trials* 1985;6:306–317.

Dickersin K, Min Y-I, Meinert CL: Factors influencing publication of research results: follow-up of applications submitted to two institutional review boards. *JAMA* 1992;267:374–378.

Doubilet P, Begg CB, Weinstein MC, Braun P, McNeil BJ: Probabilistic sensitivity analysis using Monte Carlo simulation: a practical approach. *Med Decis Making* 1985;5:157–177.

Doubilet P, Weinstein MC, McNeil BJ: Use and misuse of the term "cost effective" in medicine. *N Engl J Med* 1986;314:253–256.

Drummond M, Stoddart G, Torrance G: *Methods of Economic Evaluation of Health Care Programmes.* Oxford, Oxford University Press, 1987.

Drummond MF: Resource allocation decisions in health care: a role for quality of life assessments? *J Chron Dis* 1987;40:605–616.

Dudley HAF: Surgical research: master or servant? *Am J Surg* 1978;135:458–460.

Dupont WD, Page DL: Menopausal estrogen replacement therapy and breast cancer. *Arch Intern Med* 1991;151:67–72.

Eaker ED, Castelli WP: Coronary heart disease and its risk factors among women in the Framingham Study. In: *Coronary Heart Disease in Women.* Eaker E, Packard B, Wenger NK, Clarkson TB, Tyroler HA (eds). New York, Haymarket Doyma, 1987, pp 122–132.

Early Breast Cancer Trialists' Collaborative Group: Effects of adjuvant tamoxifen and of cytotoxic therapy on mortality in early breast cancer. *N Engl J Med* 1988;319:1681–1692.

Easterbrook PJ, Berlin JA, Gopalan R, Matthews DR: Publication bias in research. *Lancet* 1991;337:867–872.

Eddy DM: The confidence profile method: a Bayesian method for assessing health technologies. *Oper Res* 1989;37:210–228.

Eddy DM, Hasselblad V, Shachter R: A Bayesian method for synthesizing evidence: the confidence profile method. *Int J Tech Assess Health Care* 1990A;6:31–55.

Eddy DM, Hasselblad V, Shachter R: An introduction to a Bayesian method for meta-analysis: the confidence profile method. *Med Decis Making* 1990B;10:15–23.

Elbourne D, Oakley A, Chalmers I: Social and psychological support during pregnancy. In: *Effective Care in Pregnancy and Childbirth,* vol 1, *Pregnancy.* Chalmers I, Enkin M, Keirse MJN (eds). Oxford, Oxford University Press, 1989, pp 221–236.

Eysenck HJ: An exercise in mega-silliness. *Am Psychol* 1978;33:517.

Fineberg HV, Scadden D, Goldman: Care of patients with a low probability of acute myocardial infarction: cost effectiveness of alternatives to coronary-care-unit admissions. *N Engl J Med* 1984;310:1301–1307.

Finkler SA: The distinction between cost and charges. *Ann Intern Med* 1982;96:102–109.

Fisher RA: *Statistical Methods for Research Workers,* 4th ed. London, Oliver and Boyd, 1932.

Fleiss JL, Gross AJ: Meta-analysis in epidemiology, with special reference to studies of the association between exposure to environmental tobacco smoke and lung cancer: a critique. *J Clin Epidemiol* 1991;44:127–139.

Froberg DG, Kane RL: Methodology for measuring health-state preferences: I. Measurement strategies. *J Clin Epidemiol* 1989A;42:345–354.

Froberg DG, Kane RL: Methodology for measuring health-state preferences: II. Scaling methods. *J Clin Epidemiol* 1989B;42:459–471.

Froberg DG, Kane RL: Methodology for measuring health-state preferences: III. Population and context effects. *J Clin Epidemiol* 1989C;42:485–592.

Froberg DG, Kane RL: Methodology for measuring health-state preferences: IV. Progress and a research agenda. *J Clin Epidemiol* 1989D;42:675–685.

Galbraith RF: A note on graphical presentation of estimated odds ratios from several clinical trials. *Stat Med* 1988;7:889–894.

Garfinkel L, Auerbach O, Joubert L: Involuntary smoking and lung cancer: a case-control study. *J Natl Cancer Inst* 1985;75:463–469.

Garner TI, Dardis R: Cost-effectiveness analysis of end-stage renal disease treatments. *Med Care* 1987;25:25–34.

GISSI (Gruppo Italiano per lo Studio Della Streptochiasi Nell'infarto Miocardico): Effectiveness of intravenous thrombolytic treatment in acute myocardial infarction. *Lancet* 1986;1:397–402.

Gladen BC, Rogan WJ: On graphing rate ratios. *Am J Epidemiol* 1983;118:905–908.

Glass GV: Primary, secondary and meta-analysis of research. *Educ Res* 1976;5:3–8.

Glass GV, McGaw B, Smith ML: *Meta-analysis in Social Research.* Beverly Hills, Calif, Sage Publications, 1981.

Goldman L, Loscalzo A: Fate of cardiology research originally published in abstract form. *N Engl J Med* 1980;303:255–259.

Goldman L, Sia STB, Cook EF, Rutherford JD, Weinstein MC: Costs and effectiveness of routine therapy with long-term beta-adrenergic antagonists after acute myocardial infarction. *N Engl J Med* 1988;319:152–157.

Goldman L, Weinstein MC, Goldman PA, Williams LW: Cost-effectiveness of HMG-Co-A reductase inhibition for primary and secondary prevention of coronary heart disease. *JAMA* 1991;265:1145–1151.

Grady D, Rubin S, Petitti DB, Fox C, Black D, Ettinger B, Ernster V, Cummings SR. Hormone therapy to prevent disease and prolong life in postmenopausal women. *Ann Intern Med* 1992;117:1016–1037.

Greenland S: Quantitative methods in the review of epidemiologic literature. *Epidemiol Rev* 1987;9:1–30.

Greenland S, Longnecker MP: Methods for trend estimation from summarized dose-response data, with applications to meta-analysis. *Am J Epidemiol* 1992;135:1301–1309.

Greenland S, Salvan A: Bias in the one-step method for pooling study results. *Stat Med* 1990;9:247–252.

Greenland S, Schlesselman JJ, Criqui MH: The fallacy of employing standardized regression coefficients and correlations as measures of effect. *Am J Epidemiol* 1987;123:203–208.

Harris J: QALYfying the value of life. *J Medical Ethics* 1987;13:117–123.

Hebert JR, Miller DR: Plotting and discussion of rate ratios and relative risk estimates. *J Clin Epidemiol* 1989;3:289–290.

Heckerling PS, Verp MS: Amniocentesis or chorionic villus sampling for prenatal genetic testing: a decision analysis. *J Clin Epidemiol* 1991;44:657–670.

Hedges LV: Estimation of effect size from a series of independent experiments. *Psychol Bull* 1982;92:490–499.

Hedges LV: Combining independent estimators in research synthesis. *Br J Math Stat Psychol* 1983;36:123–131.

Hedges LV: Estimation of effect size under nonrandom sampling: the effects of censoring studies yielding statistically insignificant mean differences. *J Educ Stat* 1984;9:61–85.

Hedges LV, Olkin I: Vote-counting methods in research synthesis. *Psychol Bull* 1980;88:359–369.

Hedges LV, Olkin I: *Statistical Methods for Meta-Analysis.* Orlando, Florida, Academic Press, 1985.

Henschke UK, Flehinger BJ: Decision theory in cancer therapy. *Cancer* 1967;20:1819–1826.

Hillner BE, Smith TJ: Efficacy and cost effectiveness of adjuvant chemotherapy in women with node-negative breast cancer: a decision-analysis model. *N Engl J Med* 1991;324:160–168.

Himel HN, Liberati A, Gelber RD, Chalmers TC: Adjuvant chemotherapy for breast cancer: a pooled estimate based on published randomized control trials. *JAMA* 1986;256:1148–1159.

Hunter JE, Schmidt FL, Jackson BG: *Meta-Analysis: Cumulating Research Findings Across Studies.* Beverly Hills, Calif, Sage Publications, 1982.

ISIS-1 (First International Study of Infarct Survival) Collaborative Group: Randomised trial of intravenous atenolol among 16,027 cases of suspected acute myocardial infarction: ISIS-1. *Lancet* 1986;2:57–66.

Iyengar SI, Greenhouse JB: Selection models and the file drawer problem. *Stat Science* 1988;3:109–117.

Jenicek M: Meta-analysis in medicine: where we are and where we want to go. *J Clin Epidemiol* 1989;42:35–44.

Johannesson M: On the discounting of gained life-years in cost-effectiveness analysis. *Int J Tech Assess Health Care* 1992;8:359–364.

Jordan TJ, Lewit EM, Montgomery RL, Reichman LB: Isoniazid as preventive therapy in HIV-infected intravenous drug abusers: a decision analysis. *JAMA* 1991;265:2987–2991.

Kaplan RM, Ganiats TG: QALYs: their ethical implications. *JAMA* 1990;264;2503.

Kassirer JP: The principles of clinical decision making: an introduction to decision analysis. *Yale J Biol Med* 1976;49:149–164.

Kassirer JP, Pauker SG: The toss-up. *N Engl J Med* 1981;305:1467–1469.

Keeler EB, Cretin S: Discounting of life-saving and other non-monetary benefits. *Management Science* 1983;29:300–306.

Kinosian BP, Eisenberg JM: Cutting into cholesterol: cost-effective alternatives for treating hypercholesterolemia. *JAMA* 1988;259:2249–2254.

Kleinbaum DG, Kupper LL, Morgenstern H: *Epidemiologic Research: Principles and Quantitative Methods.* Belmont, Calif, Lifetime Learning, 1982.

Kuban KCK, Leviton A, Krishnamoorthy KD, Brown ER, Teele RL, Baglivo JA, Sullivan KF, Huff KR, White S, Cleveland RH, Elizabeth NA, Spritzer KL, Skouteli HN, Cayea P, Epstein MF: Neonatal hemorrhage and phenobarbital. *Pediatrics* 1986;77:443–450.

L'Abbe KA, Detsky AS, O'Rourke K: Meta-analysis in clinical research. *Ann Intern Med* 1987;107:224–233.

Laird NM, Mosteller F: Some statistical methods for combining experimental results. *Int J Tech Assess Health Care* 1990;6:5–30.

Lam TH, Kung IT, Wong CM, Lam WK, Kleevens JW, Saw D, Hsu D, Seneviratne S, Lam SY, Lo KK, Chan WC: Smoking, passive smoking and histologic types in lung cancer in Hong Kong Chinese women. *Br J Cancer* 1987;6:673–678.

LaPuma J, Lawlor EF: Quality-adjusted life-years: ethical implications for physicians and policymakers. *JAMA* 1990;263:2917–2921.

Law MR, Frost CD, Wald NJ: Analysis of data from trials of salt reduction. *Br Med J* 1991;302:819–824.

Ledley RS, Lusted LB: Reasoning foundations of medical diagnosis: Symbolic logic, probability, and value theory aid our understanding of how physicians reason. *Science* 1959;130:9–21.

Levey AS, Lau J, Pauker SG, Kassirer JP: Idiopathic nephrotic syndrome: puncturing the biopsy myth. *Ann Intern Med* 1987;107:697–713.

Light RJ (ed): *Evaluation Studies. Review Annual,* vol 8. Beverly Hills, Calif, Sage Publications, 1983.

Light RJ, Pillemer DB: *Summing Up: The Science of Reviewing Research.* Cambridge, Mass, Harvard University Press, 1984.

Lipid Research Clinics Program: The Lipid Research Clinics Coronary Primary Prevention Trial results: I. Reduction in incidence of coronary heart disease. *JAMA* 1984A;251:351–364.

Lipid Research Clinics Program: The Lipid Research Clinics Coronary Primary Prevention Trial results: II. The relationship of reduction in incidence of coronary heart disease to cholesterol lowering. *JAMA* 1984B;25:365–374.

Littenberg B: Aminophylline treatment in severe, acute asthma: a meta-analysis. *JAMA* 1988;259:1678–1684.

Llewellyn-Thomas H, Sutherland HJ, Tibshirani R, Ciampi A, Till JE, Boyd NF: The measurement of patients' values in medicine. *Med Decis Making* 1982;2:449–462.

Llewellyn-Thomas H, Sutherland HJ, Tibshirani R, Ciampi A, Till JE, Boyd NF: Describing health states: methodologic issues in obtaining values for health states. *Med Care* 1984;22:543–552.

Longnecker MP, Berlin JA, Orza MJ, Chalmers TC: A meta-analysis of alcohol consumption in relation to risk of breast cancer. *JAMA* 1988;260:652–656.

Lusted LB: Decision-making studies in patient management. *N Engl J Med* 1971;284:416–424.

Mahomed K, Hytten F: Iron and folate supplementation in pregnancy. In: *Effective Care in Pregnancy and Childbirth,* vol 1, *Pregnancy.* Chalmers I, Enkin M, Keirse MJN (eds). Oxford, Oxford University Press, 1989, pp 301–317.

Mantel N: Chi-square tests with one degree of freedom: extensions of the Mantel-Haenszel procedure. *JASA* 1963;58:690–700.

Mantel N, Brown C, Byar DP: Tests for homogeneity of effect in an epidemiologic investigation. *Am J Epidemiol* 1977;106:125–129.

Mantel N, Haenszel W: Statistical aspects of the analysis of data from retrospective studies of disease. *J Natl Cancer Inst* 1959;22:719–748.

Mast EE, Berg JL, Hanrahan LP, Wassell JT, Davis JP: Risk factors for measles in a previously vaccinated population and cost-effectiveness of revaccination strategies. *JAMA* 1990;264:2529–2533.

McCormick MC, Holmes JH: Publication of research presented at the pediatric meetings: change in selection. *AJDC* 1985;139:122–126.

Meier P: Commentary on "Why do we need systematic overviews of randomized trials?" *Stat Med* 1987;6:329–331.

Meinert CL: *Trials: Design, Conduct, and Analysis.* Oxford, Oxford University Press, 1986.

Meranze J, Ellison N, Greenhow DE: Publications resulting from anesthesia meeting abstracts. *Anesth Analg* 1982;61:445–448.

Miller GA: The magical number seven plus or minus two: some limits on our capacity to process information. *Psychol Rev* 1956;63:81–97.

Miller JN, Colditz GA, Mosteller F: How study design affects outcomes in comparisons of therapy: II. Surgical. *Stat Med* 1989;8:455–466.

Morgenstern H, Greenland S: Graphing ratio measures of effect. *J Clin Epidemiol* 1990;6:539–542.

Orwin RG: A fail-safe *N* for effect size in meta-analysis. *J Educ Stat* 1983;8:157–159.

Oster G, Epstein AM: Cost-effectiveness of antihyperlipemic therapy in the prevention of coronary heart disease. *JAMA* 1987;258:2381–2387.

Oster G, Tuden RL, Colditz GA: A cost-effectiveness analysis of prophylaxis against deep-vein thrombosis in major orthopedic surgery. *JAMA* 1987;257:203–208.

Pauker SG, Kassirer JP: Decision analysis. *N Engl J Med* 1987;316:250–258.

Pearson ES: The probability integral transformation for testing goodness of fit and combining independent tests of significance. *Biometrika* 1938;30:134–148.

Petitti DB, Perlman JA, Sidney S: Noncontraceptive estrogens and mortality: long-term follow-up of women in the Walnut Creek Study. *Obstet Gynecol* 1987;70:289–293.

Peto R: Why do we need systematic overviews of randomized trials? *Stat Med* 1987;6:233–240.

Phillips K: The use of meta-analysis in technology assessment: a meta-analysis of the enzyme immunosorbent assay human immunodeficiency virus antibody test. *J Clin Epidemiol* 1991;44:925–931.

Pliskin JS, Shepard DS, Weinstein MC: Utility functions for life years and health status. *Oper Res* 1980;28:206–224.

Prentice RL, Thomas DB: On the epidemiology of oral contraceptives and disease. *Adv Cancer Res* 1986;49:285–401.

Radhakrishna S: Combination of results from several 2 × 2 contingency tables. *Biometrics* 1965;21:86–98.

Ransohoff DF, Gracie WA, Wolfenson LB, Neuhauser D: Prophylactic cholecystectomy or expectant management for silent gallstones. *Ann Intern Med* 1983;99:199–204.

Rao CR: Comment on "Selection models and the file drawer problem." *Stat Science* 1988;3:131–132.

Rawles J: Castigating QALYs. *J Medical Ethics* 1989;15:143–147.

Read JL, Quinn RJ, Berwick DM, Fineberg HV, Weinstein MC: Preferences for health outcomes: comparison of assessment methods. *Med Decis Making* 1984;4:315–329.

Rifkin RD: Classical statistical considerations in medical decision models. *Med Decis Making* 1983;3:197–214.

Robins J, Greenland S, Breslow NE: A general estimator for the variance of the Mantel-Haenszel odds ratio. *Amer J Epidemiol* 1986;124:719–723.

Rosenthal R: Combining results of independent studies. *Psychol Bull* 1978;85:185–193.

Rosenthal R: The "file drawer problem" and tolerance for null results. *Psychol Bull* 1979;86:638–641.

Rosenthal R, Rubin DB: Further meta-analytic procedures for assessing cognitive gender differences. *J Educ Psychol* 1982;74:708–712.

Rothman KJ: *Modern Epidemiology.* Boston, Little, Brown and Company, 1986.

Sacks HS, Berrier J, Reitman D, Ancona-Berk VA, Chalmers TC: Meta-analysis of randomized controlled trials. *N Engl J Med* 1987;316:450–455.

Sacks HS, Chalmers TC, Smith H Jr: Sensitivity and specificity of clinical trials: randomized v. historical controls. *Ann Intern Med* 1983;143:753–755.

Schulman KA, Lynn LA, Glick HA, Eisenberg JM: Cost effectiveness of low-dose zidovudine therapy for asymptomatic patients with human immunodeficiency virus (HIV) infection. *Ann Intern Med* 1991;114:798–802.

Schwartz WB, Gorry GA, Kassirer JP, Essig A: Decision analysis and clinical judgment. *Am J Med* 1973;55:459–472.

Shipp M, Croughan-Minihane MS, Petitti DB, Washington AE: Estimation of the break-even point for smoking cessation programs in pregnancy. *Am J Public Health* 1992;82:383–390.

Sillero-Arenas M, Delgado-Rodriguez M, Rodriguez-Canteras R, Bueno-Cavanillas A, Galvez-Vargas R: Menopausal hormone replacement therapy and breast cancer: a meta-analysis. *Obstet Gynecol* 1992;79:286–294.

Simes JR: Treatment selection for cancer patients: application of statistical decision theory to the treatment of advanced ovarian cancer. *J Chron Dis* 1985;38:171–186.

Simes JR: Publication bias: the case for an international registry of trials. *J Clin Oncol* 1986;4:1529–1541.

Simes RJ: Confronting publication bias: a cohort design for meta-analysis. *Stat Med* 1987;6:11–29.

Sisk JE, Riegelman RK: Cost effectiveness of vaccination against pneumococcal pneumonia: an update. *Ann Int Med* 1986;104:79–86.

Smith A: Qualms about QALYs. *Lancet* 1987;i:1134–1136.

Sox HC, Blatt MA, Higgins MC, Marton KI: *Medical Decision Making.* Stoneham, Mass, Butterworth, 1988.

Spitzer WO: State of science 1986: quality of life and functional status as target variables for research. *J Chron Dis* 1987;40:465–471.

Stampfer MJ, Colditz GA: Estrogen replacement therapy and coronary heart disease: a quantitative assessment of the epidemiologic evidence. *Prev Med* 1991;20:47–63.

Stampfer MJ, Goldhaber SZ, Yusuf S, Peto R, Hennekens CH: Effect of intravenous streptokinase on acute myocardial infarction: pooled results from randomized trials. *N Engl J Med* 1982;307:1180–1182.

Steinberg KK, Thacker SB, Smith SJ, Stroup DF, Zack MM, Flanders WD, Berkelman RL: A meta-analysis of the effect of estrogen replacement therapy on the risk of breast cancer. *JAMA* 1991;265:1985–1990.

Sterling TD: Publication decisions and their possible effects on inferences drawn from tests of significance—or vice versa. *JASA* 1959;54:30–34.

Stock WA, Okun MA, Haring MJ, Miller W, Kinney C, Ceurvorst RW: Rigor in data synthesis: a case study of reliability in meta-analysis. *Educ Res* 1982;11:10–14, 20.

Sugden R, Williams A: *The Principles of Practical Cost-Benefit Analysis.* Oxford, Oxford University Press, 1990.

Thompson SG, Pocock SJ: Can meta-analysis be trusted? *Lancet* 1991;338:1127–1130.

Tippett LHC: *The Methods of Statistics.* London, Williams and Norgate, 1931.

Torrance GW: Social preferences for health states: an empirical evaluation of three measurement techniques. *Socio-Econ Plan Sci* 1976;10:129–136.

Torrance GW: Preferences for health states: a review of measurement methods. *Meade Johnson Symposium on Perinatal and Developmental Medicine* 1982;20:37–45.

Torrance GW: Measurement of health state utilities for economic appraisal: a review. *J Health Econ* 1986;5:1–30.

Torrance GW: Utility approach to measuring health-related quality of life. *J Chron Dis* 1987;40:593–600.

Torrance GW, Boyle MH, Horwood SP: Application of multi-attribute utility theory to measure social preferences for health states. *Oper Res* 1982;30:1043–1069.

Torrance GW, Thomas WH, Sackett DL: A utility maximization model for evaluation of health care programs. *Health Services Res* 1972;7:118–133.

Tosteson ANA, Rosenthal DI, Melton LJ III, Weinstein MC: Cost effectiveness of screening perimenopausal white women for osteoporosis: bone densitometry and hormone replacement therapy. *Ann Intern Med* 1990;113:594–603.

Tversky A, Kahneman D: The framing of decisions and the psychology of choice. *Science* 1981;211:453–458.

Udvarhelyi IS, Colditz GA, Rai A, Epstein AM: Cost-effectiveness and cost-benefit analyses in the medical literature: are the methods being used correctly? *Ann Intern Med* 1992;116:238–244.

United States Environmental Protection Agency: *Health Effects of Passive Smoking: Assessment of Lung Cancer in Adults and Respiratory Disorders in Children.* US EPA Publication No. EPA-600-90-006A, Washington DC, 1990.

United States Department of Health and Human Services: *The Health Effects of Smoking*

Cessation. Public Health Service, Office on Smoking and Health. DHHS Publication No. (CDC) 90-8416, 1990, pp 449–453.

Von Neumann J, Morgenstern O: *Theory of Games and Economic Theory.* New York, Wiley, 1947.

Wachter KW: Disturbed by meta-analysis? *Science* 1988;241:1407–1408.

Walker AM, Martin-Moreno JM, Artalejo FR: Odd man out: a graphical approach to meta-analysis. *Am J Public Health* 1988;78:961–966.

Ware JE Jr: Standards for validating health measures: definition and content. *J Chron Dis* 1987;40:473–480.

Warner KE, Luce BR: *Cost-Benefit and Cost-Effectiveness Analysis in Health Care: Principles, Practice, and Potential.* Ann Arbor, Mich, Health Administration Press, 1982.

Weinstein MC, Fineberg HV: *Clinical Decision Analysis.* Philadelphia, WB Saunders Company, 1980.

Weinstein MC, Stason WB: Foundations of cost-effectiveness analysis for health and medical practices. *N Engl J Med* 1977;296:716–721.

Weinstein MC, Stason WB: Cost-effectiveness of interventions to prevent or treat coronary heart disease. *Ann Rev Public Health* 1985;6:41–63.

Welch GH, Larson EB: Cost effectiveness of bone marrow transplantation in acute nonlymphocytic leukemia. *N Engl J Med* 1989;321:807–812.

Williams A: Economics of coronary artery bypass grafting. *Br Med J* 1985;291:326–329.

Williamson DF, Parker RA, Kendrick JS: The box plot: a simple visual method to interpret data. *Ann Intern Med* 1989;110:916–921.

Wilson PW, Garrison RJ, Castelli WP: Postmenopausal estrogen use, cigarette smoking, and cardiovascular morbidity in women over 50: the Framingham Study. *N Engl J Med* 1985;313:1044–1049.

Wolf FM: *Meta-Analysis: Quantitative Methods for Research Synthesis.* Newbury Park, Calif, Sage Publications, 1986.

Wortman PM, Yeatman WH: Synthesis of results of controlled trials of coronary artery bypass graft surgery. In: *Evaluation Studies. Review Annual,* vol 8. Light R (ed). Beverly Hills, Calif, Sage Publications, 1983, pp 536–557.

Yudkin PL, Ellison GW, Ghezzi A, Goodkin DE, Hughes RAC, McPherson K, Mertin J, Milanese C: Overview of azathioprine treatment in multiple sclerosis. *Lancet* 1991;338:1051–1055.

Yusuf S, Collins R, Peto R, Furberg C, Stampfer MJ, Goldhaber SZ, Hennekens CH: Intravenous and intracoronary fibrinolytic therapy in acute myocardial infarction: overview of results on mortality, reinfarction and side-effects from 33 randomized controlled trials. *Eur Heart J* 1985A;6:556–585.

Yusuf S, Peto R, Lewis J, Collins R, Sleight P: Beta blockade during and after myocardial infarction: an overview of the randomized trials. *Prog Cardiovasc Dis* 1985B;27:335–371.

Subject Index

Index of Examples